W. H. Ashmead

Orange Insects

A treatise on the injurious and beneficial insects found on the orange trees of

Florida

W. H. Ashmead

Orange Insects
A treatise on the injurious and beneficial insects found on the orange trees of Florida

ISBN/EAN: 9783742802910

Manufactured in Europe, USA, Canada, Australia, Japa

Cover: Foto ©Klaus-Uwe Gerhardt /pixelio.de

Manufactured and distributed by brebook publishing software
(www.brebook.com)

W. H. Ashmead

Orange Insects

ORANGE INSECTS;

A TREATISE

ON THE

𝕴𝖓𝖏𝖚𝖗𝖎𝖔𝖚𝖘 𝖆𝖓𝖉 𝕭𝖊𝖓𝖊𝖋𝖎𝖈𝖎𝖆𝖑 𝕴𝖓𝖘𝖊𝖈𝖙𝖘

FOUND ON THE

Orange Trees of Florida.

BY
WILLIAM H. ASHMEAD.

"Tu ne cede malis."

JACKSONVILLE, FLA.:
ASHMEAD BROS., BOOKSELLERS, STATIONERS, PRINTERS, AND BINDERS.
1880.

INTRODUCTORY.

To him who, in the love of Nature, holds
Communion with her visible forms, she speaks
A various language.—BRYANT.

IN Ornithology there are certain birds found in tropical countries, which on account of their gorgeous plumage, are called
" Birds of Paradise ; " and in fruiticulture, the orange, the
lemon, and the citron, with equal propriety, might be termed
" The Fruits of Paradise." Among the varied fruits grown or
cultivated in any clime, what equals them in beauty, in form, or
in flavor ?

The orange tree itself, with its symmetrical shape, its evergreen
glossy leaves, its flowers, surpassing in fragrance the whole Flora
Kingdom—with its golden fruit hanging in the bright sunlight,
or softened by the mild silvery rays of the moon, excells any
tree known.

Whence came this delectable fruit, fit fruit for the gods ?
What was its origin and history ? Was it stolen, by some myth-
ological hero from Elysium, like the fire by Prometheus ?

Gallesio, in his celebrated work on *The Citrus Family*, (page
43,) gives the following long but intensely interesting account of
its native country, origin, and introduction into Europe, which I
quote at length :

" The orange and lemon tree were unknown to the Romans, therefore they
could only have been indigenous in a country where this great people had never
penetrated. We all know the vast extent of this empire, yet commercial relations extend themselves always far beyond political bounds. If these trees had
been cultivated in places open to the traffic of the Romans, these fruits would
have become at once the delight of the tables of Rome, given up to luxury.
They could not, then, have been cultivated at this period, except in the remote
parts of India, beyond the Ganges. The North of Europe and of Asia, it is
true, were equally unknown to the Romans, but their climates were not at all
suited to these plants. The interior and west coasts of Africa, although in great
part deserts, and destitute of the moisture necessary to the orange, enclosed,
nevertheless, fertile districts where it might have thriven. But the state of culture of the tree at the present time in that country, and the historic facts proving

to us that it was not naturalized there till long after, make us certain t hat it was entirely unknown there as well as in Europe. It is true, that at the time of the discovery of the Cape of Good Hope, the Portugese found many citrons and bi-garades upon the eastern coast of Africa, and in the part of Ethiopia where Ro-mans had never penetrated; but they found these trees only in gardens, and in a state of domesticity, and we do not know but that the Arabs, who had cultivated them in Egypt, in Syria, and in Barbary, had penetrated into these countries in the first years of their conquests. There remains, then, for us only to seek the native country of the orange in Southern Asia—that is to say, in those vast coun-tries known under the general name of East Indies But these regions were, in part known to the Romans, who, since the discovery of the monsoons, made by Hippalus, carried their maritime commerce as far as Muziro, (Massera ; an island off the southeast coast of Arabia.—TRANS.) by way of the Red Sea, the naviga-tion of which employed a great number of vessels, and whose commerce, accord-ing to Pliny, should have been valued at fifty million sesterces ($2,000,000, T.) per annum. Their fleets had penetrated even to Portum Gebenitarum, which ap-pears to have been the present Ceylon; and, although these voyages cost them five years of fatigue and danger, nevertheless, the thirst for gold and luxury of Rome had multiplied to the last degree the vessels engaged in this trade. We must believe, then, that the lemon and orange did not exist in all that part of the country this side of the Indus, and perhaps not even in all the part lying between that river and the Ganges; otherwise these fruits would have been extolled by the Roman merchants—where the citron was so much valued; and we should find at least some mention made of them in narratives and voyages descended to us from those ancient times. If we consult the description of the coasts of India—from the river Indus to the Euphrates, which we have in the voyage of Nearchus, one of Alexander's captains; that of the Troglodytes, and coasts of the Indian Sea, by Arianus, the voyage of Iambolus, reported by Diodorous of Sicily, where he gives a description of an isle of the Indian Sea unknown before him, where he had been thrown by a storm ; or, finally, the Indian voyage of Pliny—we find not the least indication of either orange, or even citron ; yet. Nearchus carefully notes the plants found in his course, and speaks of palms, myrtles and vines ; of wheat; and generally of all the trees of Asia except the olive. Arianus enlarges upon the vegetable productions of those districts giving the descriptions of those found in public roads.

Iambolus saw in the unknown island, which appears to have been Sumatra, a grain that we recognize as maize, which has been introduced into Europe since the passage round the Cape of Good Hope. We must then admit that the lemon and orange trees could not have originated but in the region beyond the Ganges, and that, in early centuries of the empires of the Cæsars, they had not yet been brought from those climates where they were indigenous. They increased per-haps still without culture in the midst of the woods, the hand of man not having yet appropriated them as ornaments for his garden. But this event could not long be delayed. The beauty of the tree, and the facility 'with which it repro-duced itself, would natuarally extend the culture to adjoining provinces, and the European, quick to seize the productions of all the rest of the globe, would not fail to enrich himself from these regions.

Facts prove that this result has been reached. but we know not the date of pas-sage, or the circumstances favoring it. We will now make this the object of our researches. The Romans, at the time of Pliny, had extended their commerce on the side of India, as far as it was ever carried during the empire; the power of Rome instead of increasing, only became weaker from this period; and the fall of the Western portion was accompanied in Europe by the decay of letters, arts, agriculture and commerce. In this general overturn, the Greeks preserved, it is true, with a taste for arts and luxury, some relations with India, but trade with those countries had never taken other course than by way of the Red Sea, and this was closed from the seventh century by the Arabian invasion of Egypt, which soon followed the invasion of Arabia by the Barbarians of the west. (Ethiopians T.)

The commerce of these rich lands must then have taken a much longer and more dangerous route. The traders were obliged, after going down the Indus, to reascend that stream; and by the Bactrea (Bolkh) to reach the Oxus—and finally, by the last pass into the Caspean Sea, from whence they went into the Black Sea by the river Don. But this long and dangerous voyage was never undertaken by the traders of Constantinople; they would not have been able to traverse with safety such an extent of country partly a desert, and in part inhabited by wandering tribes, most of them nations with whom they were nearly always at war; who were destined in the end to swallow the Greek Empire.

They therefore limited themselves to receiving upon the borders of the Caspean Sea, the merchandize of India, brought to them by intermediate people. One can scarcely realize that in such a state of affairs the orange tree could pass into Europe, for this beautiful part of the world had never been in so general disorder or had so little intercourse with India. Her luxury and commerce were nearly annihilated, and the Arabians, whom the new religion of Mahomet rendered fanatics and conquerers, menaced on one side the tottering empire of the Greeks, and on the other threatened to plunge into barbarism the West, just beginning to be civilized. Yet it was precisely at this point of time, and by the conquering spirit of this people that the great changes were prepared which should revive and extend further than ever before the commercial relations of Europe with Asia, and of Asia herself with the more distant regions of her own continent.

The Arabs, placed in a country which binds together three grand divisions of the globe, have extended their conquests into Asia and Africa much farther than any people before them. Masters of the Red Sea, and Mediterranean, they had invaded all the African coast this side of Atlas; and penetrated beyond to the region of the Troglodytes, (Ethiopians living in caves—Trans.,) the ancient limit of the Roman establishments on the east coast of this continent; they had made settlements there, and according to the testimony of a historian of the country, cited by Barros, they had populated in the fourth century of the Hegira, (A. D. 944;) the towns of Brava, Mombas and Quiloa, whence they extended themselves to Sofalo; Melinda and to the islands of Bemba, Zanzibar, Monfra, Comoro and St. Laurent. On the side of Asia they had carried their conquests in the third century of the Hegira, to the extremities of the Reinahar, and towards the middle of the fourth century, under the Selucidæ, they had established a colony at Kashgar, the usual route of caravans to Toorkistan or to China, and which, according to Albufeda, (a geographer and historian of Damascus, Trans.,) is situated in longitude 87 deg. (78 deg., 57 min. Trans.,) consequently they had penetrated very far into Asia.

Never had there been in Asia an empire so vast, and never had the commerce of nations so near Europe been pushed so far into India.

A position thus advantageous and favorable to the commercial spirit and love of luxury which succeeded, among the Arabs, the fury of conquest, would naturally cause them to learn of, and to appropriate many exotic plants peculiar to the regions they had conquered, or to the adjoining countries. Fond of medicine and agriculture, in which they have specially excelled, and of the pleasures of the open country, in which they have always delighted, they continued to profit with eagerness from the advantages offered by their settlements, and the hot climates which they inhabited. Indeed, it is to them that we owe the knowledge of many plants, perfumes, and Oriental aromatics, such as musk, nutmegs, mace and cloves.

It was the Arabs who natualized in Spain, Sardinia, and Sicily, the cotton-tree of Africa, and the sugar-cane of India: and in their medicines we for the first time hear of the chemical change known as distillation, which appears to have originated in the desire to steal from nature the perfumes of flowers and aroma of fruits.

It is then not surprising that we are indebted to them for the acclimatization of the orange and lemon trees in Syria, Africa and some European islands. It is certain that the orange was known to their physicians from the commencement of the fourth century of the Hegira. The Damascene has given in his Antidotary, the recipe for making oil with oranges, and their seeds, (OLEUM DE CITRANGULA, ET OLEUM DE CITRANGULORUM SEMINIBUS. MAT. SILV., F. 58,) and Avicenna, who died in 428 of the Hegira, (1050) has added the juice of the bigarade to his syrup of ALKEDERE ET SUCCI ACETOSITATIS CITRI (otrodj,) ET SUCCI ACETOSITATIS CITRANGULI, (narendg)." These two Arabians seem to have first employed it in medicine.

I have examined with care the authors of this nation who preceded these, and find in no other the least hint relating to these species. Mesue, even, who speaks of the citron, says not a word of orange or lemon. I have observed, on the contrary, that Avicenna, in giving his recipe for making syrup of alkedere, in which he puts juice of the bigarade, announces it as a composition of his own invention. This circumstance would indicate that this fruit had been known but a short time in Persia, but it suffices that it was cultivated there to prove that it might, at once, pass into Irak, (probably Irak-Arabee, in Asiatic Turkey, comprising Bagdad, TRANS.) and into Syria.

These countries which joined, were also connected by political ties, which facilitate communication, and their inhabitants were more civilized then, than before or since. A passage by Massoudi, reported by the learned M. de Sacy in the notes to his translation of Abd-Allatif, a writer of the twelfth century of our era, seems to confirm our ideas upon this subject, and to determine the date of this event. It accords with all the data just given, and with historic facts that we have collected. He expresses himself thus: "The round citron (otrodj modawar) was brought from India since the year 300 of the Hegira. It was first sowed in Oman (part of Arabia, TRANS.) from thence carried to Irak, (part of Old Persia, TRANS.) and Syria, becoming very common in the houses of Tarsus and other frontier cities of Syria, at Antioch, upon the coasts of Syria, in Palestine and in Egypt. One knew it not before, but it lost much of the sweet odor and fine color which it had in India, because it had not the same climate, soil, and all that which is peculiar to that country." The lemon appeared perhaps a little later in these different countries, for we see no mention of it either in the Damascene or in Avicenna, but its description meets our eye in all the works of Arabian writers of the twelfth century, especially Ebn-Beitar, who has given to it an article in his dictionary of simple remedies. The Latin translation of this article was published in Paris in 1702 by Andres Balunense. The Imperial library contains several manuscripts of this dictionary.

I had thought to have found proof that the lemon was known by the Arabs in the ninth century, having seen in a history of India and China, dated 238 of the Hegira, (A. D. 860. T.) of which a French translation was printed in Paris in 1718, the writers had spoken of the lemon as a fruit found in China.   But M. de Sacy who examined the original, ascertained that the word limon, was inserted by the translator. In the Arabian text one finds only that of OTRODJ. which signifies merely OITRON.   Therefore this history, far from proving that the Arabs knew the lemon tree at this period, proves quite the contrary.   It was not until the tenth century of our era that this warlike people enriched with these trees the garden of Oman, (in southeastern Arabia, Trans.) whence they were propagated in Palestine and Egypt.   From these countries they passed into Barbary and Spain, perhaps also into Sicily.

Leon of Ostia, tells us that in 1002, a prince of Salerna presented citrine apples (POMA OITRINA) to the Norman princes who had rescued him from the Saracens.

The expression POMO OITRINA. used by this author appears to me to designate fruit like the citron rather than the citron itself, then known under the name of OITRI, or of MALA MEDICA.

It is thus that we should recognize the orange in the OITRON ROND spoken of by Massoudi in a passage already quoted.   This conjecture accorded with known events and data.   The Arabs invaded Sicily about the beginning of the ninth century, (828,) the orange was taken from India to Arabia after the year 300 of the Hegira—that is to say, early in the ninth century of our era.   The citrine apples of Leon d' Ostia dates from 1002, and were regarded as objects rare and precious enough to be offered as gifts to princes.   Thus we have between its introduction into Arabia, and propagation in Sicily an interval of nearly a century.   In order to conform to the expression of Massoudi, let us suppose that the orange tree was brought from Arabia some thirty or forty years later—say about 330 of Hegira, if we allow fifty years for its propagation in Palestine, Egypt and Barbary, and finally twenty years for its naturalization in Sicily, we fill precisely the interval between one epoch and the other.

A passage in the history of Sicily, by Nicolas Specialis, written in the fourteenth century, gives still more probability to this opinion.

This writer, in recounting the devastation by the army of the Duke of Calabria in 1383, in the vicinity of Palermo, says that it did not spare even the trees of sour apples, (pommes acides;) called by the people arangi, which had adorned since old times, the royal palace of Cubba.   (Nicolas Specialis, bk. 7, c. 17.)

The name Cubba given to this royal pleasure house, seems to refer to the time of the Arabic rule, it is probably derived from the Arabic word Cobbah, meaning vault or arch, perhaps some grand dome upon this country-house gave the place its name.

These data, however, do not appear to me sufficiently strong to combat the authority of a very reliable historian, who says expressly that the lemon and the orange trees were not known in Italy or France or in other parts of Christian Europe in the eleventh century.   Such are the words of Jacques de Vitry, in speaking of Syrian trees in his history of Jerusalem.   The testimony of this bishop, who ought to have known these countries, would appear to have more weight than simple conjectures based upon reasonings from analogy.   Whatever be the authority of this historian, compared with the presumptions advanced by us with regard to Sicily, it will always be decisive respecting Lake Garda and the coasts of Liguria and Provence.

There is not a doubt that in these last named countries the lemon and orange were unknown; not only in the tenth but even in the eleventh century. But an extraordinary event, destined to change the face of Europe, was to open anew to the people of the West the entrance to Syria and Palestine. This was also the time when the Crusades, which began at the close of the eleventh century (1096; Tr.), reawakened among Europeans the spirit of commerce and a taste for arts and luxury.

The Crusaders entered Asia Minor as conquerors, and thence spread themselves as traders into all parts of Asia. They were not mere soldiers, but brave men drawn from their families by religious enthusiasm, and who, in consequence, would hold fast to their country and their homes. They could not see without coveting these charming trees which embellished the vicinity of Jerusalem, with whose exquisite fruits Nature has favored the climates of Asia.

It was, indeed, at this time that Europe enriched its orchards by many of these trees, and that the French princes carried into their country the damson, the St. Catharine (a pair, Tr.), the apricot, from Alexandria, and other species indigenous to those regions.

Sicilians, Genoese and Provincials transported to Salermo, St. Remo and Hyeres the lemon and orange trees. Hear what a historian of the thirteenth century says to us on this subject; he had been in Palestine with the Crusaders, and his word should have great weight—

Jacques de Vitry expressed himself thus—"Besides many trees cultivated in Italy, Genoa, France, and other parts of Europe, we find here (in Palestine) species peculiar to the country, and of which some are sterile and others bear fruit: Here are trees bearing very beautiful apples—the color of the citron—upon which is distinctly seen the mark of a man's tooth. This has given them the common name of pomme d' Adam (Adam's apple)—others produce sour fruit, of a disagreeable taste (pontici), which are called limons. Their juice is used for seasoning food, because it is cool, pricks the palate, and provokes appetite.. We also see cedars of Lebanon, very fine and tall, but sterile. There is a species of cedar called cedre maritime, whose plant is small but productive, giving very fine fruits —as large as a man's head. Some call them citrons or pommes citrines. These fruits are formed of a triple substance, and have three different tastes. The first is warm, the second is temperate, the last is cold. Some say that this is the fruit of which God commanded in Leviticus—'Take you the first day of the year the fruit of the finest tree.' We see in this country another species of citrine apples, borne by small trees, and of which the cool part is less and of a disagreeable and acid taste—these the natives call orenges."

Behold, then, the Adam's apple, the lemon, the citron, and the bigarade found in Palestine by the Crusaders, and regarded as new trees foreign to Europe.

This passage does not accord, as far as the citron is concerned, with what Palladius says. He tells us that this plant was, in his time, cultivated in Sardinia and in Sicily. But we see, by Jacques de Vitry, that the citron of Palestine was distinguished by the extraordinary size of its fruit, equal to a man's head, and it must be that this last was a variety unknown to Europe.

It is, indeed, only since this epoch that we find in European historians and writers upon agriculture any mention of these trees. Doubtless the Arabians had already naturalized them in Africa and Spain, where the temperature favored so much their growth. Doubtless Liguria is the part of Italy where the culture of the Agrumi has made most progress. We have certain testimony to this in the work of a doctor of medicine of Mantua, writing near the middle of the thirteenth century. He says—

"The lemon is one of the species of citrine apples, which are four in number. First, citron. Secondly, orange, (citrangulum,) of which we have spoken before. Thirdly, the lemon. Fourthly, the fruit vulgarly called lima. These four species are very well known, principally in Liguria. The lemon is a handsome fruit, of fine odor. Its form is more oblong than that of the orange, and, like the orange, it is full of a sharp, acid juice, very proper for seasoning meats. They make of its flowers odoriferous waters, fit for the use of the luxurious.

"The trees of these four species are very similar and all are thorned. The leaves of the citron and lime are larger and less deeply colored than those of the orange or lemon. The lemon is composed of four different substances, as well as the citron, lime and orange. It has an outer skin, not as deep in color as that of the orange, but which has more of the white. It is hot and biting, thus it shows its bitter taste. The second skin or pith, between the outer skin and the juice, is white, cold, and difficult to digest. The third substance is its juice, which is sharp, and of a strong acid, which will expel worms, and is very cold. The fourth is the seed, which, like that of the orange, is warm, dry and bitter." (See Mat. Silv., Pandecta Medicinæ, fol. 125.)

This testimony of Silvaticus is strengthened by all the authors who have written upon the citrus. There is not one but is convinced that these trees were for a long time very rare in Italy and in France, and that Liguria alone has traded in them since they were first known there. Sicily and the kingdom of Naples cultivated, perhaps before the Ligurians, the citron and orange trees; but in spite of the advantage of climate, it was only, as objects of curiosity, limited to some delightful spots. This fact is established by the manner in which most writers of the twelfth century express themselves on this subject. Hugo Falcandus, who wrote of the exploits of the Normans in Sicily, from 1145 to 1169, saw there lumies and orangers, and points them out as singular plants, whose culture was still very rare. (Hugo Falcandus. See Muratori, Rerum Italicarum Scriptores.)

Ebn-Al-Awam, an Arabian writer upon agriculture at Seville, near the end of the twelfth century, and whose work, translated into Spanish, was published at Madrid in 1802, speaks as if the culture were very much extended in Spain. Abd-Allatif, who was cotemporary with the last-named author, expresses himself in like manner, and describes also a number of varieties cultivated in his time in Egypt—a circumstance showing that these trees had greatly multiplied. Their progress was slower in Italy and France. It appears that the lemon tree, brought first into these parts as a variety of citron, was for a long time designated by European writers under the generic name of citrus, although in Italy and the South of France the people had known it from the beginning under the proper name of limon—a name which has come down to us without submitting to any change. In fact, we find it in botanical works called citrus limon, or mala limonia, and sometimes citrus medica. The last was indefinitely used to designate lemon, citron and orange, and very often the genus citrus.

The orange appeared in Italy under the name of orenges, which the people modified according to the pronunciations of the different sections, into arangio, naranzo, aranza, aranzo, citrone, cetrangolo, melarancio, melangolo, arancio. One meets successively all these names in works of the thirteenth, fourteenth and fifteenth centuries, such as those of Hugo Falcandus. Nicolas Specialis, Blondus Flavius, Sir Brunetto Latini, Ciriffo Calvaneo, Bencivenni, Bocaccio, Giustiniani,

Leandro Alberti, and several others. The Provencals also received this tree under the name of orenges, and have changed it from time to time in different provinces, into arrangi, airange, orenge and finally orange. (See Glossary of the Roman language, by Roquefort.)

During several centuries the Latin authors found themselves embarrassed in designating this fruit, which had no name in that language. The first who spoke of it used a phrase indicating its characteristics, accompanying it with the popular name of arangi, latinized into orenges, arangias, arantium.

Thus, Jacques de Vitry, who calls the oranges poma citrina, adds, "The Arabs call them orenges." And Nicolas Specialis designated them as pommes aigres (acripomorum arbores), observing that the people call them arangias. These have been followed by Blondus Flavius and many others. Matheus Silvaticus first gave to the orange the name of citrangulum, and this denomination seems to have been followed for a long time by physicians and translators of Arabic works, who have very generally adopted it for rendering the Arabic word, arindj.

Thus, citrangulum was received for more than a century in the language of science. Finally, little by little, were adopted the vulgar Latinized names in use among other writers, such as authors of chronicles, etc., and they have written successively, arangium, arancium, arantium, anarantium, nerantium, aurantium, pomen aureum. The Greeks followed in the same steps. They have either Grecianized the name of narenge, which was in use among Syrian Arabs, or they received it from the Crusaders from the Holy Land, and have adopted it in their language, calling it nerantzion. These have, however, always been considered vulgar names, and, in general, the better Latin writers have made use of the generic name, citrus, for designating the Agrumi.

This usage, followed by most of the writers on history and chorography, often occasions uncertainty and difficulty in researches concerning the beginning of this culture in the different countries where these trees have been introduced. The use of it as seasoning for food, brought from Palestine to Liguria, to Provence, and to Sicily, penetrated to the interior of Italy and France. The taste for confections was propagated in Europe with the introduction of sugar, and this delicate food became at once a necessary article to men in easy circumstances, and a luxury upon all tables. It was, above all, as confections, that the Agrumi entered into commerce, and we see by the records of Savona that they were sent into cold parts of Italy, where people were very greedy for them.

After having cultivated these species for the use made of their fruits, they soon cultivated them as ornaments for the gardens. The monks began to fill with these trees the courts of their monasteries, in climates suited to their continual growth, and soon one found no convent not surrounded by them. Indeed, the courts and gardens of these houses show us now trees of great age, and it is said that the old tree, of which we see now a rejeton in the court of the convent of St. Sabina, at Rome, was planted by St. Dominic, about the year 1200. This fact has no other foundation than tradition, but this tradition, preserved for many centuries, not only among the monks of the convent, but also among the clergy of Rome, is reported by Augustin Gallo, who, in 1559, speaks of this orange as a tree existing since time immemorial. If we refuse to attribute its planting to St. Dominic, we must at least refer it to a period soon after, that is, to the end of the thirteenth century, at the latest.

Nicholas Specialis, in the passage cited on another page, in describing the havoc made by the besiegers in the suburbs of Palermo, regrets the destruction of orangers, or trees of sour apples (pommes aigres), which he regards as rare plants, embellishing the pleasure-house of Cubba.

Blondus Flavius, a writer of the middle of the following century, speaks of the orange on the coast of Amalfi (a city of Naples, Tr.) as a new plant, which as yet had no name in scientific language, (Blond-Flav., Ital. Illus., p. 420), and he extols the valleys of Rapallo and San Remo, in Liguria, for the culture of the citrus, a rare tree in Italy. Cugus ager (San Remo,) these are his words, est citri, palmaquæ, arborum in Italia rarissirarum, ferax. (Blond. Flav., Ital. Illust., p. 296.) Lastly, Pierre de Crescenzi, Senator of Bologna, who wrote in 1300 a treatise on agriculture, speaks only of the citron tree. We find in his expressions no hint of lemon or orange. The culture of these trees, then, had been begun in the fourteenth century only in a few places, but was extended in proportion as arts and luxury advanced the civilization of Europe.

The orange was from the first valued not alone for the beauty of its foliage and quality of its fruit, of which the juice was used in medicine, but also for the aroma of its flowers, of which essences were made. Pharmacists have employed with success the juice of the lemon in making medicines.

The orange tree must have been taken to Provence about the time it entered Liguria. It is to be presumed that the city of Hyeras, so celebrated for the softness of its climate and the fertility of its soil, received it from the Crusaders, because from this port the expeditions to the Holy Land took their departure. We see, indeed, that it was greatly multiplied there, and in 1566 the plantations of oranges within its territory were so extensive and well-grown as to present the aspect of a forest.

The territory of Nice, so advantageously placed between Liguria and Provence, would necessarily receive from its neighbors a tree so suited to the softness of its climate, sheltered by the Alps, and to the nature of its soil, fertilized by abundant waters. It appears that the culture had already greatly extended towards the middle of the fourteenth century, as we find in the history of Dauphiny that the Dauphin Humbert, returning from Naples in 1336, bought at Nice twenty plants of orange trees. (Hist. of Dauphiny, bk. 2, p. 271.)

From Naples and Sicily the orange and lemon trees must have been carried into the Roman States into Sardinia and Corsica and to Malta. The islands of the Archipelago perhaps first received them, because, belonging in great part to the Genoese and Venetians, it is probable they were the intermediate points whence the Crusaders of Genoa and Venice transported the plants to their homes. From these isles the trees have afterwards spread into the delightful coast of Salo on the shores of Lake Garda, where, in Gallo's time (1559,) they were regarded as acclimated from time immemorial. Finally, the orange and the lemon penetrated into the colder latitudes, and perhaps one owes to the desire of enjoying their flowers and fruit the invention of hot-houses, afterwards called orangeries. (The name of orangerie is a modern word in the French language. Olivier de Serre does not use it—he calls this kind of inclosure orange-houses, p. 633. The Italian language has no word responding precisely to orangery. We find in some modern authors, equivalent words, such as aranciera, cedroniera, citronera. Fontana, Dizionario rustico, bk. 1, p. 74. But the ancient writers styled these places for preserving these trees by the phrase, Stanzone per i cedri. In Tuscany and the Roman States, they call them rimesse. In other places they are known under the name of serre (inclosure). Matioli says, that in his time they cultivated the oranges in Italy on the shores of the sea and of the most famous lakes, as well as in the gardens of the interior, but he says nothing of the places for sheltering them. Gallo speaks of rooms designed to receive the boxes of orange-trees, which were very numerous at Brescia, but he does not designate them by any particular name. The Latin writers also used a periphrase. Ferraris calls an orangery, tectum hibernum. Others call it cella citraria.)

This agricultural luxury was unknown in Europe before the introduction of the citron-tree. We find not the least trace of it either in Greek or Latin writers.

It is true that from the time of the Emperor Tiberius in Rome, they enclosed melons in certain portable boxes of wood, which were exposed to the sun in winter to make the fruit grow out of season. These inclosures were secured from the effects of cold by sashes or frames, and received the sun's rays through diaphanous stones, (specularia,) which held the place of our glass. But it seems they used no fire for heating them, and that they merely inclosed thus indigenous plants, of which they wished to force the fruiting out of season, it being a speculation of the cultivator rather than a luxurious ornament for embellishing the gardens. (Pliny, bk. 19, chap. 5, p. 336, and Columell, bk. 2, chap. 8, p. 42.) It is after the introduction of the citron tree into Europe that we begin to find among the ancients examples of artificial coverings and shelters against cold. Palladius is the first who speaks of these coverings, but only as appropriate for the citron, and gives no description of them. Florentin, who wrote probably after him, describes them at more length, and it seems by his expressions, that in his time the citron was covered in the bad season by wooden roofs, which could be withdrawn when there was no occasion to defend them from cold, and which, also, could be arranged to secure for them the rays of the sun. (Florent. bk. 10, chap. 7, p. 219.)

This agricultural luxury, which began to appear about the time of Palladius and Florentin, must have been entirely destroyed in Italy by the invasion of the barbarians. I have remarked that Pierre de Crescenti, who wrote a treatise on agriculture in 1300, while treating of the citron, speaks only of walls to defend it from the north, and of some covers of straw. Brunsius and Antonius, quoted by Sprengel, have thought to find in the Statutes of Charlemagne indications of a hothouse. I have closely examined the article cited by those writers, (in Comment. de reb. Franc. orient, bk. 2, p. 902, etc.,) but have not found a word that could make me believe this means of preserving delicate plants was employed at that period. I have even remarked that in these ordinances many plants are named which Charlemagne wished to have in his fields, but no word to be construed into ordering a shelter for any, unless the fig and almond. It is astonishing that having spoken in detail of all the parts of the house, of laboring utensils the most ordinary—and even of those of housekeeping—he forgot an object of such great luxury as a hot-house. But in proportion as civilization and commerce increased riches and extravagance, the fruit of this tree became more sought for, and at the same time, more common ; whilst, above all, the properties of the new species just introduced extended its use in medicine, in agreeable drinks, and as a luxury of the table. At first they were in cold countries only a foreign production procured from the South, but afterwards the people began to covet from the more happy climates the ornament of these trees, and to wish, above all, to embellish with them their gardens. In temperate climes they began to cultivate them in vases, depositing them during winter in caves; and in the cold latitudes the necessity of struggling against nature, gave the idea of constructing apartments which could be heated at pleasure by fire, and which would shelter the plants from the rigor of the season.

It is difficult to fix the date at which they began to build edifices for protection of oranges. The oldest trace of it that I have been able to find, is furnished by a passage in the history of Dauphiny, dated 1336, (we find in this history, printed at Geneva in 1722, an extract from an account of expenses made by Humbert, the Dauphin, in his voyage of Naples in 1336. In the expenses for the return we see the sum of ten tarins—the tarin was the thirtieth part of an ounce of Naples—for the purchase of twenty orange plants. Item pro arboribus viginti de plantis aran. giorum ad plantandum taren. X. Hist. of Daup., bk. 2, p. 276.) This, it is true, offers few circumstantial details for fixing the fact that the princes of Dauphiny had really, at that time, an orangery; but as this historian tells us that Humbert bought at Nice twenty roots of orangers for a plantation, (ad plantandum,) it is to be supposed that he had in his palace at Vienna a place designed to preserve them in the winter; for, without this precaution, they certainly would have perished in the rigorous climate of Dauphiny. (In southwest part of France. Tr.)

This luxury must have passed immediately into the capital of France, and though I have not yet found in history indications of these establishments before 1500, it is very probable that they were known there about the middle of the four-teenth century.

The celebrated tree, preserved still in the orangery at Versailles, under the name of Francis First, or Grand Bourbon, was taken from the Constable of Bourbon, in the seizure made of his goods in 1523. And this prince, who, it is said, possessed it for eighty years, could not have kept it except in an orangery. (The orange tree at Versailles, known as Francis Premier, is the most beautiful tree that I have seen in a box. It is twenty feet high, and extends its branches to a circumference of forty feet. Spite of that I scarcely believe that this fine stalk dates from the fourteenth century. It is too vigorous, and the skin is too smooth, to be able to count so many years. It is probable that in so long a course of time it has been cut, and that the present tree is a sprout from the old root. This might have oc-curred after the frost of 1709, which penetrated even into sheltered places. One circumstance gives foundation to this conjecture. This tree is composed of two stalks, which both come out of the earth, and have a common stock. This is never the way the tree grows by nature, still less in a state of culture, and from roots held in vases. I have mostly remarked it in the greater number of trees growing upon a stump which had been razeed at the level of the ground. In such case one is forced to leave two suckers, because the sap, being very abundant, could not develop itself in one shoot. It would experience a sort of reaction which would suffocate the stump and make it perish. This is a well-known fact in the South, where we cultivate largely the orange, and where the trees of double stems are generally recognized as rejetons, or suckers from old roots.)

After all these data, we are authorized to think that in the fourteenth century they had begun already to erect buildings designed to create for exotic plants an artificial climate. But at the beginning of the fifteenth century orangeries passed from kings' gardens to those of the people, chiefly in countries where they were not compelled to heat them by fire, as in Brescia, Romagna and Tuscany. (See Matioli, who says that in his day the orange was cultivated in Italy, in all the gardens of the interior, where certainly it could not live, unless in orangeries. Diosc. c. 132. We also find in Sprengel's History of Botany, that in this country there were at that time many botanical gardens where they cultivated exotic plants—a circumstance which presupposes the necessity of hot-houses.)

About the middle of the seventeenth century this luxury was very general, and we see distinguished by their magnificence and grandeur, the orangeries of the Farnese family at Parma, of the Cardinals Xantes, Aldobrandini and Pio, at Rome, of the Elector Palatin at Heidelberg, (Oliv. de Ser., p. 633,) of Louis Thirteenth, in France; and even at Ghent, in Belgium, that of M. de Hellibusi, who imported plants from Genoa, and carried his establishment to the last degree of magnificence. (See Ferraris, p. 150, where he describes the orangery of M. de Hellibusi at Ghent, and that of Louis Thirteenth at Paris. The latter has been replaced by that of Versailles, of which the magnificence renders it perhaps the finest monument of this kind to be found in Europe.)

We now see orangeries in all the civilized parts of Europe, it being an embellishment necessary to all country-seats and houses of pleasure.

THE ORANGE IN FLORIDA.

From all accounts the orange is not indigenous to Florida, although vast groves of wild orange trees exist in different parts of the State. It was nevertheless brought here by the Spaniards probably in the beginning of their conquests, in the early part of the sixteenth century.

The wild orange has a bitter, acrid taste, and according to some, agrees with that grown in modern days in Spain, known as the "Seville Orange." Whatever its origin in Florida by cultivation, budding and importation, it has become greatly changed, and the "Florida Orange" for its size, beauty and flavor, has gained a reputation second to none in existence.

On coming to Florida for the first time in the winter of 1876, I found *orange culture* the great industry of the State. Go where I would, orange groves and their culture was the principal theme of conversation. Indeed, the "orange fever" was contagious, and very few visitors to the "Land of Flowers" escaped without imbibing some of its contagion. However, on investigation, I found there was good cause for the enthusiasm.

When it is known that a small grove of one hundred trees on an acre of land, *in full* bearing condition, yields annually an income of from a thousand to two thousand dollars, or a grove of ten acres, an income of ten thousand dollars or more—is it to be wondered at, that people become so wild and enthusiastic over their culture? There is a tree near Palatka that produces annually twelve thousand oranges.

Another thing I discovered about the orange tree was that like the vegetables of the garden, the trees of the orchard, the apple, peach, pear, &c., it had different kinds of insect enemies, that

"Myriads on myriads
Keen in the poison'd breeze would wasteful eat
Through bud and bark into the blackened core.".

which lived on the foliage, fed on the sap, or otherwise proved destructive. I also learned that they in their turn were preyed upon by others.

Among the most injurious of these were various kinds of *Bark-Lice* or *Scale Insects*, (*Coccidæ*,) which frequently killed the tree or damaged it to such an extent as to cause whole branches to die back, and otherwise retarded the growth and vigor of the tree. The majority of these I found had been imported from foreign shores, and had been taken up

"On the wing of the heavy gales
Through the boundless arch of heaven."

to lodge again on trees in other groves, until they were distributed to all parts of the State.

Thousands of dollars are annually lost to orange-growers from the depredations of these pests. Having always had a "hobby" for bugs and " curious creatures " of every kind, I immediately set to work reading up, studying and investigating their habits. Within the past two years, I have received so many communications from parties all over the State, requesting information in regard to these insects, and feeling that the publication of my studies and researches would fill a want long needed by the orange-grower, I have concluded to publish them in book-form, laying no claim to literary merit, but only desiring that they may prove valuable to the fruit-grower, by enabling him to discern friends from foes, and after learning their habits, stimulate him to investigate and experiment for successful methods for their destruction.

Having thus briefly given my reasons for bringing myself before the public, after having given a short account of the orange-tree, I will now proceed to treat of its insect inhabitants—friends and foes.

THE LONG, OR MUSSEL-SHELL SCALE.

(*Mytilaspis* [*Aspidiotus*] *Gloverii*, Packard.)

[Ord., HEMIPTERA. Fam., COCCIDÆ.]

●

BIBLIOGRAPHICAL.

Aspidiotus Gloverii, Packard, Guide to Study of Insects.

The first of the many scale insects of the orange tree to be treated of, is the Long Scale, or Mussel-Shell Scale, named by Prof. Packard, *Aspidiotus Gloverii*, after Townend Glover, recently of the Agricultural Department at Washington, D. C. According to the latest revision of the *Coccidæ* by M. Signoret, of France, it belongs properly to the genus *Mytilaspis*, but as it is generally known in Florida under the name of *Aspidiotus*, I have still, in brackets, retained that name for it.

ITS IMPORTATION AND SPREAD.

In the year 1835, (date given me,) it was first brought into this country on some orange trees imported from China, first making its appearance in the grove of Dr. Robinson, at Mandarin, a small town on the St. Johns River, about twelve miles from Jacksonville. In a few years, it had spread to the groves throughout Florida, carrying devastation and ruin wherever it went. So great was the damage that orange growers became discouraged, and orange culture was nearly annihilated. Many groves that had been yielding handsome incomes, were totally destroyed. Happily, however, in a few years the scales became less numerous, orange trees were again planted, their culture revived, and the result is seen in the rapidly growing industry. Florida, to-day, produces between fifty and sixty million oranges, and in two or three years will double that number.

ITS INTRODUCTION INTO FLORIDA.

Mr. Brown, in his "Trees of America," gives the following account of its introduction and spread:

"This insect first made its appearance in Florida, in Robinson's grove, at Mandarin, on the St. John's, in 1838, (this date is evidently a mistake,) on some trees of the Mandarin orange, which had been procured in New York. In the course of three or four years they had spread to the neighboring plantations, to the distance of ten miles, and were the most rapid in their migrations in the direction of the prevailing winds, which evidently aided them in their movements. In 1840, Mr. P. S. Smith, of St. Augustine, obtained some orange trees from Mandarin and had them planted in his front yard. From these trees the insects went to others in the same enclosure, and rapidly extended themselves to the trees and plantations to the northerly and westerly parts of that city and its vicinity, obviously aided in their migrations by the southeast trade winds, which blow there almost daily during summer; and what is remarkable, these insects were occupied nearly three years in reaching trees in the southeast side of the city, only about half a mile from their original point of attack. They have since, however, extended themselves to all the trees in and about the city, but have not yet traveled in any direction beyond two miles. Being aided in their dispersion by birds and other natural causes, impossible to guard against, they must eventually attack most if not all the trees in Florida; for the wild orange groves suffer equally with those which have been cultivated, and no difference can be perceived in their ravages between old and young trees, nor between vigorous and decayed ones.

"Various remedies have been tried to arrest their progress—such as fumigating the trees with tobacco-smoke, covering them with lime, potash, sulphur, shellac, glue, and other viscid and tenacious substances mixed with clay, quick-lime, salt, etc.; but all have failed, partially or entirely, and it appears not to be in the power of man to prevent the ravages of these insignificant and insidious destroyers. Most of the cultivated oranges in Florida have already been injured by them, their tops and branches having been mostly destroyed. Their roots and stems, it is true, remain alive, and annually send forth a crop of young shoots, only to share the fate of their predecessors. The visitation of these insects in Florida, probably, is not destined to continue much longer, at least with its present violence; for, among the means which nature has provided to check their increase, are various species of birds that devour inconceivable numbers of them, and the *coccidæ* are invariably accompanied by considerable numbers of yellow lady-bugs, (*coccinella*.) which, it has been conjectured, have been appointed to keep them down."

Many other insects are found preying upon them, which I shall describe further on.

METHOD OF SPREADING.

Various, theories in regard to the manner in which these insects spread from tree to tree and grove to grove, have been promulgated and published in different journals, particularly in the Florida papers. In my opinion, there are but three principal ways of transportation: First, on nursery stock; second, by the wind; third, on the fruit itself.

ON NURSERY STOCK.

From reliable authorities we learn that the specie under consideration, was imported into Florida in this way, and in like manner distributed to other groves, until to-day, it is, I believe, found in the United States wherever the orange is cultivated.

BY THE WINDS.

In support of this, Mr. Brown says, "From these trees, the insects went to others in the same enclosure, and rapidly extended themselves to the trees and plantations in the northerly and westerly parts of the city and its vicinity, obviously aided in their migrations by the southeast trade winds, which blow there almost daily during summer; and what is remarkable, these insects were occupied nearly three years in reaching trees in the southeast side of the city, only about half a mile from their original point of attack." Now, in the spring and fall, just when the young insects are hatching and most numerous, we have our heaviest storms, sometimes regular hurricanes blowing at the rate of forty, fifty, or more miles per hour. During one of these storms, I have often seen leaves, twigs, and sometimes whole branches taken up and carried whirling through the air for a quarter of a mile or further.

How easy, then, would it be for these microscopical insects, but a few atoms in weight, to be carried for miles. The seeds and pollen of plants and flowers, trees and shrubs, are borne for hundreds of miles through the air, from their original starting point. Would it then be any more wonderful for these small insects to be carried in like manner?

ON THE FRUIT ITSELF.

Townend Glover, in his Report for 1855, states that the oval scale, *Aspidiotus Citricola*, was imported in this manner on some lemons from Bermuda. (See account of Oval Scale, further on.)

If any one will take the trouble to examine imported oranges or lemons brought into the northern markets, he cannot fail to find scales on some of them. These oranges are sent by fruit-dealers to every part of the country, and in this manner many of the different species of insects are distributed.

ITS NATURAL HISTORY.

The elongated, dark brown scale, (Fig. 1, twig infested,) consisting of a series of successive waxy secretions, (Fig. 2, upper figure highly magnified, after Glover,) which harden from the effects of the sun and atmosphere, is from .05 to .10 of an inch in length and is nearly twice as broad posteriorly as anteriorly; it is attached either to the young twig, bark, leaves, or fruit—attacking all indiscriminately. When fully matured, under this scale are laid, in two parallel rows, thirty or more ovoid eggs about .01 of an inch in length, of a bright red, becoming purplish before hatching. (Fig. 2, lower scale.) They are somewhat contracted at smaller end, and are not full, well-formed eggs like those of other species.

Within twenty to thirty days, these hatch, and the young larvæ are then about .01 of an inch in length, of a pale yellow color, with six jointed antennæ, and devoid of caudal filaments.

Fig. 1.

They crawl for three or four days over the leaves and twigs, then finding a suitable place, insert their beaks, become stationary—never afterwards moving. In a few days the waxy covering, or shell, begins to form over them; the legs, as they have no more use for them, drop off, and the insect, by a retrograde development, changes into a legless larval form, of a pinkish or flesh color. (Plate 1, fig. 8.)

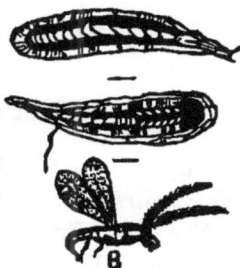

Fig. 2. (After Glover.)

This species differs very much from all others I have examined It is very elongated, with well-defined thorax, head, and abdomen. Underneath the abdomen, the different segments comprising it are furnished with claws, which enables it to crawl backwards and forwards in its long scale. Plate 1 (Fig. 7, side view, greatly enlarged) gives a good idea of these. When it has reached this stage it is fully matured, and soon afterwards lays its eggs. The male, (Fig. 2,) unlike the female, is furnished with two wings, which enables it to migrate wherever it pleases. It is of a reddish color, with long, hairy, ten-jointed antennæ and black eyes; abdomen paler, and furnished with a long, curved penis; it has no beak; the place where it should be is indicated by two or three black dots; consequently it never feeds, but soon after performing the duties for which it was created, dies.

EXPLANATION TO PLATE I.

Figs. 1 and 2.—Female and winged individual of *Siphoniphora citrifolii*. ASHMEAD.

Fig. 3.—Wings of same, showing venation.

Fig. 6.—Beak of same.

Fig. 4.— Female *Trichogramma flavus*. ASHMEAD.

Fig. 5.—Hind legs and coxæ of same.

Fig. 7.—Side view of larval form of *Mytilaspis Gloverii*, showing claws on segments.

Fig. 8.—Larval form of same, as it appears with scale removed.

Fig. 9.—Larval form of *Aspidiotus lemonii ?*

Fig. 10.—Larval form of *Lecanium hesperidum*, with hind legs gone, and showing transparent spot with viscera in centre. In life the round dot palpitates up and down.

Fig. 11.—Egg of *A. lemonii*.

Fig. 12.—Scale of *L. hesperidum*.

Figs. 13 and 14.—Showing nervous and digestive systems of *hesperidum*, after Lubbock.

Fig. 14.—*gg*, hepatic glands; *w*, œsophagus, (long and narrow;) *f*, pyriform gland, crop or stomach, with a remarkable cellular contorted internal gland; *d*, ilium, short intestine, opening into the rectum, (c;) *b*, narrow tube leading into rectum; *hh*, recurrent intestine, two ends of which are attached to the stomach, (f;) *e*, cœcum, swelling at its base, and is perhaps the equivalent of the sucking stomach.

Fig. 13.—Ganglionic column; *c*, large nervous column, with two divisions, each of which again subdivides, giving a rich plexus of nerves to posterior part of body; *b*, nerve, throwing off an inner nerve, (f.)

PLATE I.

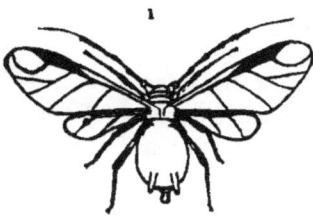

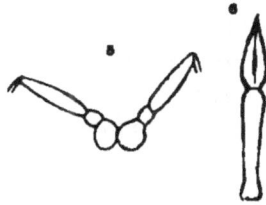

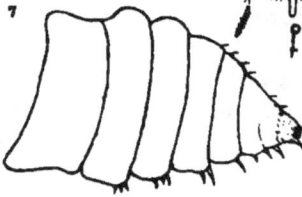

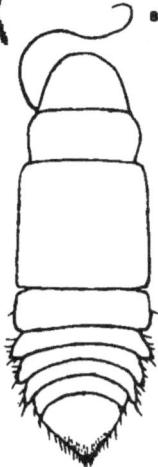

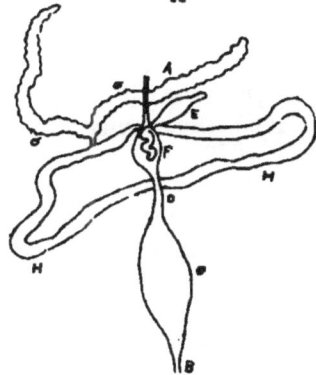

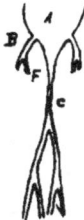

ITS NATURAL ENEMIES.

One cause of its rapid spread soon after its introduction, is probably due to the fact that its natural enemies (species of chalcid flies) were not imported with it; consequently it had full sway, and nothing to prevent its prolific breeding.

In time, some of our own native insects began preying upon it, and to these is probably due the fact of their ceasing to be so destructive as they were the first few years of their importation.

The following are the principal ones so far discovered, an account of which will be given in regular order, according to to their importance:

THE ORANGE SCALE APHELINUS: Aphelinus aspidioticola, Ashmead.

THE TWICE-STABBED LADY BUG: Chilocorus bivulnerus, Muls.

THE MINUTE SCYMNUS: Hyperaspidius coccidivora, N. Sp.

THE RED MITE OF THE ORANGE: Oribates Gloverii, Ashmead.

THE ORANGE CHRYSOPA: Chrysopa citri, N. Sp.

THE ORANGE SCALE APHELINUS.

(*Aphelinus aspidioticola.* Ashmead.)

[Ord., HYMENOPTERA. Fam., CHALCIDIDÆ.]

BIBLIOGRAPHICAL.

Aphelinus aspidioticola, Canadian Entomologist, Vol. XI.
Aphelinus aspidioticola, Florida Agriculturist, Vol. II, No. 17, 1879.

"A friend in need is a friend indeed."

It was but a few years after the orange groves of Florida had been blasted and ruined by scale insects, that this wonderful and welcomed foe to these pernicious pests made its appearance. It is one of the chief agencies in holding them in check, and I am firmly satisfied could these minute foes be removed but for a season, the numerous scales would so increase in numbers as to utterly destroy all orange trees, and another panic in orange culture would ensue as disastrous and uncontrollable as the one witnessed in the years from 1835 to 1840. How important, then is it for us to study the life-histories of these destructive insects—to seek out their habits, find their likes and dislikes, discover their foes, and thus by their means, if by no other, control and keep them in subjection to our will. What a glorious science, then, is Entomology to the fruit-grower, agriculturist, and florist!

ITS FIRST APPEARANCE.

The first account we have of our little friend is that given by Glover, in his Report on Orange Insects, in the United States Agricultural Report for the year 1855. He says, "Another hymenopterous fly came out of the dead scales, which also measured about the twentieth part of an inch in length, the thorax and first segment of the body being light brown, with the rest of the abdomen blackish and hairy; the head was furnished with three ocelli; the four wings were transparent, and the antennæ long, jointed, and hairy."—(Page 119.) I described it in the Canadian Entomologist, Vol. XI, under the name of *Aphelinus aspidioticola*—i. e., inhabitant of the *aspidiotus*.

ITS NATURAL HISTORY.

It is about .02 of an inch in length; of a light brownish color, with four wings, ciliated, and agrees very much with the description of the Aphelinus of the apple-tree A. mytilaspis, found preying upon the apple scale-insect, Mytilaspis pomicorticis, (Riley,) but differs in the following respects: it is smaller, the abdomen is considerably longer than thorax, with the wings ciliated but from just before the apex; antennæ, too, is different. (Plate 2, fig. 1, gives a good idea of it.) It lays its eggs, a single one, under each scale, among the eggs of the scale insect. On hatching, the larva, which is a white, fleshy, footless grub, immediately begins feeding upon them. After it has destroyed all the eggs, and has reached full growth, it changes into a pupa, (plate 2, fig. 1;) remaining in this condition for a few days, it then transforms into the four-winged fly, as described above, making its exit from under the scale by eating a round hole in the top.

Groves badly infested with scales, in which these flies do not appear, I would recommend the transportation of Aphelini into them. Le Baron, State Entomologist of Illinois, successfully transported the Aphelinus of the apple scale, A. mytilaspis, into apple orchards, and the experiment proved not only successful but beneficial. It can easily be accomplished by taking branches infested with scales, that are known to have been chalcidized, and tying them on the infested trees.

DESCRIPTIVE.

APHELINUS ASPIDIOTICOLA. FEMALE — Length about .02 of an inch. Head wider than thorax, both light reddish brown—head nearly same width as thorax, three ocelli forming a triangle—compound eyes, prominent, dark—antennæ eight-jointed, setaceous, first joint longer than 2, 3, 4 and 5, second joint round, nearly twice as wide as third, other joints gradually increasing in size, somewhat truncate anteriorly, last or apical joint large and club-shaped—legs, yellowish, long and setaceous, with a tibial spur, tarsi long, five jointed—wings, hyaline, ciliated from stigma—with numerous small short bristles on the surface—abdomen longer than thorax, upper surface of segments more or less dusky, blackish towards apex, with several hairs surrounding ovipositor.

MALE—Agrees very much with above description, excepting it is slightly smaller and scape or first joint of antennæ is shorter and broader, the other joints more rounded than in the female. with red spot on thorax at base of each wing, with upper surface of abdominal segments dark brownish.

TWICE-STABBED LADY BUG.

(*Chilocorus bivulneris*, Muls.)

[Ord., COLEOPTERA. Fam., COCCINELLIDÆ.]

This important factor in the destruction of scale insects and plant lice is quite widely spread over the United States, and its benefactions to the agriculturist and fruit-grower have been noticed and recorded by many observers.

ITS NATURAL HISTORY.

Early in the season, from February to November, in Florida, they and their dark slate-colored larvæ, which are covered with numerous spines, may be seen crawling up and down the trunk of the orange tree, in the branches or wherever the scale exists. They breed throughout the whole year. The female, late in the fall, lays her eggs wherever the scales are thickest; early in the spring the small spiny larvæ hatch, and immediately begin feeding upon them. On reaching maturity they crawl off to a retired place, suspend themselves from a leaf or twig, and change into pupæ, transforming in a few days into beetles, which make their exit from the pupa-skin by a longitudinal slit down the back. (Fig. 3.) On emerging, the beetle is soft and of a pale color, without signs of spots, but within a short time the elytra harden, color darkens to black, two red spots on wing-covers appear, and the perfect insect (fig. 4) is before us. Should there be any Spanish moss (*Tillandsia usneoides*) on the tree, the larvæ will invariably congregate, and transform attached to it. As they are very important in destroying the scale insects, every care should be taken to increase their numbers.

Fig. 3.

I would also recommend transporting the larvæ into groves where they do not exist, as it could be done without much expense, with but little trouble, and prove of incalculable value. Two dozen, or more, could be sent in a small tin box by mail for two or three cents.

Fig. 4.

Several instances of the larvæ being mistaken for injurious insects have been brought to my notice; there is now no excuse for such mistakes, as the admirable cuts will acquaint any one with its various forms.

THE MINUTE SCYMNUS.

(*Hyperaspidius coccidivora*, N. S.)

[Ord., COLEOPTERA.   Fam., COCCINELLIDÆ.]

This small beetle, on account of its diminutiveness, has never before been noticed by the orange grower. I first detected it with its larvæ in April, 1878, along with *Scymnus cervicalis*, and the spiny larva of the Twice-stabbed Lady Bug, (*Chilocorus bivulnerus*,) running up the trunks of orange-trees, near Jack-

sonville. I sent specimens of it, with other Coleoptera, to Dr.
GEO. H. HORN, of Philadelphia, for determination, and he reported it new to science. I shall, therefore, designate and describe it under the name *H. coccidivora.*

ITS NATURAL HISTORY.

By looking carefully in April and May, the beetle (fig. 5) can easily be distinguished on the trunks of the trees, with its larva, which is flattened and of a uniform brownish color.

Although so small and insignificant, it accomplishes Fig. 5 a great deal in the destruction of the scale insects. The larvæ, hatching in the spring, at the same time with the young scale insects, immediately begin their warfare upon them, which they continue even after they have transformed into beetles.

It would be difficult to estimate the benefit derived from these little beetles. After May and June, the majority of them suddenly disappear, and we do not see them again until the great fall brood of scales hatch.

Through the kindness of Dr. GEO. H. HORN, I am enabled to give the following description:

HYPERASPIDIUS COCCIDIVORA, N. S.—Broadly oval, convex, piceous, shining, each elytron with a large, badly-defined rufous space, which sometimes reaches the side margin and suture. Thorax sparsely, finely punctate. Elytra more coarsely punctured. Body beneath, and legs piceous, shining. Length, .04 inch.

This insect resembles some of our smaller Scymnus, but it is entirely without pubescence. It is not larger than PENTILIA PUSILLA, and from its resemblance to that insect, except in color, would have been referred to that genus, but there are six abdominal segments.

THE LARGE SCYMNUS.
(*Scymnus servicalis*, Muls.).

[Ord., COLEOPTERA. Fam., COCCINELLIDÆ.]

This is a hemispherical beetle, of a reddish brown color, with very dark blue elytra, and from .08 to .10 of an inch in length. It is not so numerous as the previously described species, although by no means rare. It evidently helps the others in destroying the scales. The larva, which I have not seen, is quoted by Dr. Packard, as follows: " Body, subcylindrical, pale whitish, much longer, slenderer, and narrower than in Coccinella, with a small, black, round head; the legs are long and slender, more so than in Coccinella.   The rings are rather convex, not tuberculated above, though provided with a few hairs.   It is .12 of an inch long."

THE RED ORANGE-MITE.
(*Oribates aspidioti*, Ashmead.)

This mite was first described by me in "Canadian Entomologist," for May, 1879, and I have not since met with it.   It was found feeding upon the eggs of the long scale, (*Aspidiotus Gloverii*,) infesting an orange branch, brought to me by Mr. ALLEN H. CURTISS, who resides about five miles from Jacksonville.   It is, therefore, not so widely spread as *Tyroglyphus Gloverii*, to be hereafter treated of.   That it may be known when seen, I quote the original description.

"ORIBATES ASPIDIOTI.—Elongate, flattened, narrowing towards head, dark reddish brown color—abdomen, pubescent, with two oval capitate processes, the first in centre, just back of cephalothorax, the second just below middle of abdomen, and both striate—outer edge slightly serrated—four legs, stout, and with but one claw curved inwards, with three or four basil hairs.   Length, about .02 inch.

Plate 2, Fig. 11, represents another mite (length, .04 inch) that I am inclined to think is a more matured form of the above.   At first, it is reddish and changes to a dark brown, and is very hard and tough.   There are also other mites of same color as this, only the abdomen is round.   These, too, I take for immature forms.

THE ORANGE CHRYSOPA.
(*Chrysopa citri*, N. Sp.)

[Ord., NEUROPTERA.   Fam., CHRYSOPIDÆ.]

Fig. 6.

Another beneficial insect to the orange culturist, is a lace-winged fly, which I detected preying upon the scale insects two years ago; and for which I proposed the name of *Chrysopa citri*, at the last meeting of the Florida Fruit-Growers' Association. (See " Florida Dispatch," March 3, 1880.)

ITS NATURAL HISTORY.

The eggs of this species are suspended upon a delicate silver thread, nearly half an inch long upon the upper and frequently to the under part of an orange leaf. They are elongate oval, about .05 of an inch in length, and are either greenish yellow or purplish in color, according to age. (Fig. 10, plate 2.)

The larva, on hatching, covers itself with minute pieces of dried leaves and other particles; when full grown it resembles the common ant lion. (Plate 2, fig. 14.) It is pinkish, and beautifully mottled with brownish spots. The spiracles of the thoracic segments are very long, growing with the larva and to suit the rubbish on its back, smaller air-vessels ramify from the end of the spiracle tube through the mossy mass. Plate 2, fig. 4, gives a good representation of these spiracles after the mossy covering has been removed. It feeds on the scale insects and plant-lice, particularly the latter, securing them by its long curved pincer-like mandibles. It then forms an oval moss-like cocoon on the upper part of the leaf, and in a few days transforms into the perfect fly, (fig. 6,) with long antennæ, net-winged, and bright golden eyes, shining in the dark like coals of fire.

DESCRIPTIVE.

CHRYSOPA OITEI, N. Sp.—Bright yellowish green. Antennæ, light brownish longer than wings, and finely annulated, dark reddish brown at base, extending to one-third the length. Head, greenish yellow, three dark spots on occiput, two just back on neck. Wings, hyaline, iridescent; length to tip, .67 inch—veins, greenish. Length of body from head to end of abdomen, .38 of an inch.

We have also found two other chrysopa flies upon the orange tree—C. PLORABUNDA, and the other is an undetermined species of Hemerobius.

EXPLANATION TO PLATE 11.

Figure 1.—Female of *Aphelinus aspidioticola*, Ashmead.

Fig. 4.—Fore wing of same.

Fig. 7.—Antenna of same.

Fig. 9.—Hind leg.

Fig. 13.—Antenna of male.

Fig. 2.—*Signiphora flavopalliatus*, Ashmead.

Fig. 3.—Antenna.

Figs. 6 and 8.—Wings of same.

Fig. 12.—Fore leg, showing coxæ.

Fig. 15.—Hind leg, showing appendage.

Fig. 11.—Pupa of a chalcid fly found under oval scale.

Fig. 14.—Larva of *Chrysopa citri*, with mossy covering removed, showing spiracles.

Fig. 16.—Cocoon of same on orange leaf, showing method of escape of perfect fly.

Fig. 17.—Red mite of orange, supposed to be a more matured form of *Oribates aspidioti*, Ashmead.

Fig. 18.—*Phytoptus oleivorus*, Ashmead.

PLATE II.

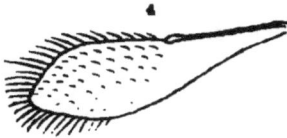
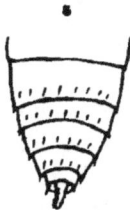

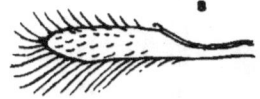

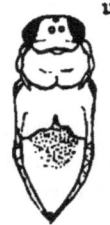

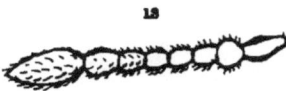

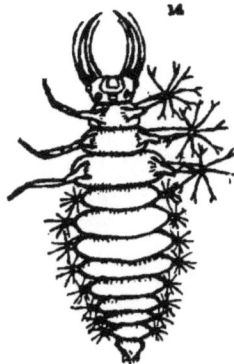
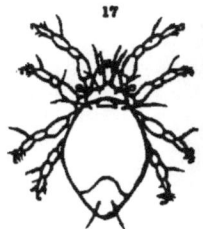

REMEDIES.

Many washes have been devised and recommended as a sure destroyer of these pests, the majority of which I believe are efficacious, if they reach the insects, which they frequently never do.

The great difficulty is to apply the wash at the proper time, when the young are soft and just hatched.

It is just here where the Entomologist steps in, and proves the usefulness of his bug-investigating propensities. The scale, supposed to be by many the insect proper, is merely a waxy-like covering, forming a protection first for the insect, and afterwards for its eggs, and, like a good roof, is impervious to all external substances. Consequently, washes applied at this stage of their existence are worthless, expensive, and labor lost. In this particular scale, as I have shown, there are three distinct broods in a season; hence there must be three seasons in which to apply the washes—i. e., just after each brood has hatched. Knowing when these hatch, the time varying but slightly from that given under the head of its natural history, in parts of Florida, from difference in climate, moisture, or backwardness in seasons, one can have no difficulty in exterminating and keeping them under control.

For trees badly affected I would recommend cutting the smaller limbs off and burning, as the most rapid method of destroying; the balance of the tree should then receive an application of one of the washes, as soon as the young scale insects hatch. The next rainy spell would soon replace the limbs removed, and afterwards one would have no trouble in keeping his trees free from scales.

Washes.—A wash made from a decoction of tobacco leaves and whale-oil soap, and syringed upon the parts infested with the young lice, will always kill them. To the trunks of trees badly affected, it would be advisable to apply the wash recommended by Dr. Harris, in his celebrated work on Insects Injurious to Vegetation, which is as follows:

"A wash made of two parts of soft soap and eight of water, with which is to be mixed lime enough to bring it to the consistence of thick whitewash. This is to be put upon the trunks and limbs of the trees with a brush, and as high as practicable, so as to cover the whole surface and fill all the cracks in the bark.

THE WHITE SCALE.

(*Ceroplastes rusci*, Linn.)

[Ord., HEMIPTERA. Fam., COCCIDÆ.]

BIBLIOGRAPHICAL.

Coccus rusci, Linn., Sys. Nat.—Fab. Syst. Ent., 1775.
Id., Spec., Ins., 1781. Id., Matis Ins., 1787.
Id., Ent. Sys., 1794. Id., Sys. Ryng, 1803.
Modeer, Act. Goth., 1778. Gmelin, Syst. Nat., 1791.
Oliv., Encyc. Meth., 1791. *C. curicæ*, Fab. Ent. Syst., 1794.
Id., Syst., Ryng., 1804. Bernard, Mem. Hist. Nat., et Mem.
Acad., 1775. Fronscol, Ann. Soc. Ent., 1834.
Boisduval, Ent., Hort., 1867. *Lopus tesselata*, Klein, 1734.
Calypticus testudineus, Costa, 1837. Faun. Regm., Nap., Gallins.
Columnea testudinata, Targioni, 1866. Atti dei, Georgof.
Id., Studi Sulle Coccinig, Ext. Soc. Ital. Scien., Milan., et Catal,
1868. (Signoret.)

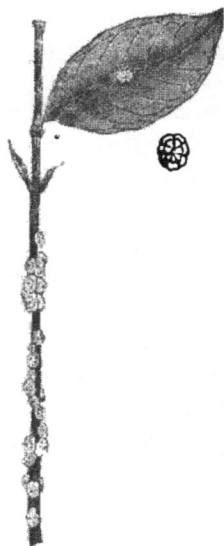

Fig. 7.

This species, as the bibliographical account shows, has been renamed by various authors, thus creating for it many synonyms.

It was first described by Linnæus, under the name of *C rusci*, and although subsequently redescribed by various authors, it must still retain the original name given by him.

ITS DISTRIBUTION

It is widely distributed, being found in Europe, Australia, and the Southern parts of America. Like the long scale, *Aspidiotus Gloverii*, it has probably been imported into this country, as I can find no record of its having been found in Florida several years back, and it is now just beginning to become common.

M. Signoret "Essai sur les Cochinelles" gives its food plants in Europe as the myrtle, common holly, and wormwood. In Florida, I have found it on the myrtle, orange, fig, and oleander.

Prof. J. H, Comstock, in his tour through Florida last spring, told me he found it in abundance on the gallberry, (*Ilex glaber*.)

The scale, (Plate 3, fig. 1, greatly enlarged; fig. 7, branch of myrtle, showing scales attached,) when fully matured, averages from .10 to .14 of an inch in length, by from .06 to .08 of an inch in breadth, and is highly arched. On the top it is tessellated with seven well-defined, oval, elevated checkers, three on each side, nearly round, the seventh at posterior, being more or less triangular.

At first the color is whitish, resembling wax, with which it is similar in consistency, being soft and pliable. As it reaches maturity, it becomes pinkish, with a slight yellowish tint in depressions. Just before the young hatch, it takes a globular form, and the top changes to a dark brown. The summer brood of young hatch from sixteen to twenty days after the egg has been laid. The female (Plate 3, fig. 4) is flattened, oval, resembling, when highly magnified, a wood-louse, only not so convex. It is pale yellow in color, with two light brownish lateral dorsal stripes, probably caused by the viscera. In the posterior end is a deep triangular indentation, from the centre of which protrudes a fleshy tubercle, reaching to outer edge; from each corner of the notch, on either side of the tubercles, is a long caudal filament, nearly as long as the body. (Plate 3, figs. 3 and 4.) On each side of these filaments are short hairs. The eyes are round and black, and can be seen from below and above. Antennæ, moderately long, six-jointed, (not easily made out,) with three long hairs on inner side, and three at apex, the inner being the longest. They crawl around for two or three days after hatching, then insert their beaks, and become attached to the surface of the leaf or bark; the waxy secretion forms over them in small globules, and in a few days is plainly visible in the form of small, white, round, elevated spots surrounding the insect, particularly just above the spiracles, (breathing-holes.) As it increases in size, the limbs, which are of no more use, gradually disappear, (Fig. 2, plate 3,) and on reaching maturity it forms a brownish oval pupa, (Plate 3, fig. 6,) which encloses its eggs, over one hundred in number. In one I counted one hundred and five ! These are elliptical, and of a pale yellow color. (See fig. 7, plate 3.)

NUMBER OF BROODS.

During the year there are three broods; the first brood hatches in April and May; second, middle of July to first of August; third, last of August to second week in September.

Some idea of their prolificness can be formed by supposing that nothing prevents the hundred eggs of the first brood from hatching out in April, and each egg being allowed to mature; in July, each insect again produces a hundred, making just ten thousand ; now, suppose these, too, hatch, each producing its hundred, this would give a grand total of one million.   One million, the offspring of a single insect in one season !   At this rate, it would not take long for every orange, fig, or quince tree in Florida to be filled with them.   Thanks, however, to that immutable law which governs the universe, they have their enemies to prey upon and keep them under.

ITS NATURAL ENEMIES.

The Twice-Stabbed Lady-Bug, *Chilocorus bivulnerus;* the Blood-Red Lady-Bug, *Cycloneda sanguinea*, and a chalcid fly are its chief enemies.   The latter, I have as yet been unable to procure specimens of, although I have found many of the scales perforated with a hole in the top, through which it made its exit. I hope, in time, to secure these and determine the species.

I have also discovered a very small, active, minute mite, (Plate 3, fig 12,) about 300ths of an inch in length, with eight legs, two pair thrust forward and two pair backwards, crawling about among the eggs and old scales.   Whether it is an enemy or not, I cannot say, as I have failed to detect it doing any injury.   I am inclined to think it merely preys upon the decaying matter of the old scales.

DESCRIPTIVE.

EGGS.—From 80 to 105 under each scale, less than 0.01 of an inch in length, elliptical, more than twice as long as broad, smooth, full and well formed—at first whitish, changing to a pale yellow before hatching, and promiscuously enclosed in dried-up body walls of the female.

LARVA, OR FEMALE.—Length about .01 of an inch, flattened, pale yellow, nearly three times as long as broad, and with two pale brownish, longitudinal dorsal bands, one on each side of the middle, probably caused by the viscera. Antennæ six jointed, third and last joint longest, other joints thick and irregular, the last ending in three or four hairs, two of which are long, interior one being the longer. Also two long inner hairs, with other shorter hairs below these. Legs, normal, ending in a feeble claw, surrounded by three or four short hairs and one digituli. Abdomen consists of seven segments, and deeply triangularly notched at the end, with a fleshy tubercle protruding from the centre, reaching to outer edge, two long anal setæ, nearly the length of body, with a short hair on either side, a few short hairs surrounding outer edge.

SCALE.—Oval, highly arched when matured, from .10 to .14 of an inch in length by from .06 to .08 of an inch in width, and nearly the same in height, tessellated on top with seven well-defined checkers, lateral three roundish, the seventh more or less triangular. Color whitish or pinkish, tinted in depressions with yellow, losing most of the checkers and becoming globular and brownish with age.

EXPLANATION TO PLATE III.

Figure 1.—Showing tesselations on Scale of *Ceroplastes rusci*.

Fig. 2.—Showing insect under scale, after losing one pair of legs.

Fig. 3.—Under surface of Scale fully matured.

Fig. 4.— Female, soon after hatching.

Fig. 5.—Same viewed from beneath.

Fig. 6.—Matured female, having lost its limbs, and with piece of body wall removed so as to show eggs enclosed.

Fig. 7.—Egg, highly magnified.

Fig. 8.—Egg, with insect hatching.

Fig. 9.—Egg Shell after escape of insect.

Fig. 10.—Antenna.

Fig. 11.—Foot, showing digituli.

Fig. 12.—Mite, found crawling under the old scale.

PLATE III.

1

2

8

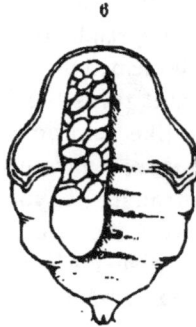

4

6

5

7

9

10

8

12

11

THE RED, OR CIRCULAR SCALE.

(*Chrysomphalus ficus*, Riley.)

[Ord., HEMIPTERA.   Fam., COCCIDÆ.]

BIBLIOGRAPHICAL.

Chrysomphalus ficus, Riley, Manuscript Notes.
Chrysomphalus ficus, Ashmead, Fla. Agriculturist, 1879.
Chrysomphalus ficus, Ashmead, Pacific Rural Press, 1880.

ITS FIRST APPEARANCE IN FLORIDA.

In September, 1879, I received the following communication, with specimens of infested leaves, from Mr. G. M. Holmes:

"ORLANDO, Orange County, Sept. 20, '79.

"WM. H. ASHMEAD—

"Dear Sir :—Enclosed I send you a leaf of an orange tree infested with what appears to be a species of scale insect, which is new to us down here.   It spreads from tree to tree very rapidly, and is not confined to the leaf, but appears upon tender stems and thorns.   You can see it turns the leaf yellow wherever it locates itself.   I should like to know whether it is an enemy much to be dreaded, and if you have had experience with it and the cure.   Although a stranger to you, I see by the Florida Agriculturist, you have made the insects on orange trees a study, and I thought you might give me some information about this particular insect.   Yours, respectfully,

G. M. HOLMES."

The scale being new to me, I immediately forwarded specimens to Prof. C. V. Riley, and from his reply I quote the following:

"The Circular, Dark Brown Scale, with a golden centre, has long been in my cabinet, and I have found it quite injurious to *ficus nitida*.   I have designated it by the manuscript name, *Chrysomphalus ficus*, but have not published any description of it, as the mere description of the scale, without fully characterizing the insect that makes it, in both sexes, is imperfect entomological work."

ITS IMPORTATION AND SPREAD.

In Los Angeles, San Jose, California, and indeed, in various parts of the State, it is quite numerous on the orange, and is

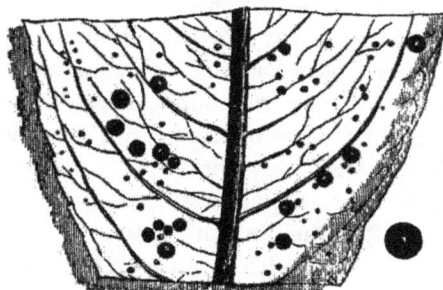

there known as the "*Red Scale.*"

The orange tree has but lately been introduced and grown in California, and this particular species is, therefore, not indigenous there.

Where, therefore, did it come from, and how was it

Fig. 8.

introduced into the State? These are two very important questions.

Now, the commercial relations existing between the Californians with the people of China, Japan and Australia, point to either one of these countries as the original home, or starting point from which it has spread. Indeed, many oranges have been imported from all these places, and it would not be surprising if, like our own *Long Scale*, the *Red Scale* had been imported in the same manner—i. e:, on the leaves, branches or twigs of an imported tree.

ITS FOOD PLANT.

Prof. C. V. Riley states he first found it on the *Ficus nitida.* This, I presume, is an exotic species of fig. I see by the Pacific Rural Press, that this, or an allied species, had been found on the apple trees in San Jose, Cal. With the orange, it attacks the fruit, leaves and twigs, seeming to like one about as well as the other.

ITS NATURAL HISTORY.

I have not been able to work up this insect thoroughly for want of specimens. What little I have done, is due to the kindness of Mr. G. M. Holmes, who has kindly sent me specimens from which my cuts and figures have been made. Figure 8 represents part of an orange leaf with scale attached.

From specimens received at different times, I know there are at least three broods, if not more. The first brood probably hatches in May; the second, from the last of July to the second week in August; and the third, last of September to first week in October; these two last broods I have raised from specimens sent me by Mr. Holmes.

The young (Plate 4, figs. 4 and 6) are less than .01 of an inch long, nearly twice as long as broad, and of a *bright* yellow color; antennæ, six-jointed, ending at tip into two long hairs, the inner being the longer, with three inner and two outer hairs lower down; the abdomen has no indentation or notches like the other species, and the two anal filaments are very short. The figures in Plate 4, give an excellent idea of the different parts, and no one can fail to recognize them after studying the figures.

The egg (Plate 4, fig. 7) is less than .01 of an inch long, and is of a bright yellow color, not quite twice as long as wide. It takes from five to six weeks for this to mature; in this time, if we examine one of the scales, we will find a footless larva, (Plate 4, fig. 5,) underneath it of a golden color. This soon dies, enclosing its eggs in its own body. The young, on hatching, force their way out after remaining two or three days under the scale. Sometimes they attach themselves to the leaf under the old scale, all huddled in a heap together. When this is the case, all die but two or three, as there is not enough sustenance to keep all alive, and the weaker must succumb to the stronger.

ITS NATURAL ENEMIES.

Being so far from the vicinity where this species is found, I am unable to give a list of its foes, but know that the Twice-Stabbed Lady Bug, and one or two other insects belonging to the *Coccinellidæ*, are found where it occurs, and must prey upon it to a certain extent.

REMEDIES.

Mr. G. M. Holmes writes me from Orlando, Fla., under date of August 6th, as follows: "As you request, I forward you by this mail a box containing specimens of the *Chrysomphalus ficus*, which I hope may reach you in good order. They have not done me any material damage as yet, but I keep my trees in a very healthy condition and thrifty growth, as I have a large drove of cattle, and can cow-pen them. In my experiments for their removal, I have been most successful in the use of a strong brine of salt and water, applied twice, at intervals of two weeks. It is heroic treatment, and takes the leaves off, but the scale comes with them, and if done just prior to a growing season, they soon send out a luxuriant new growth and seem more healthy than before   I think if potash was mixed with salt and water it would be an improvement, and I am going to use it that way. You have my best wishes for success with your book, which will supply a want much felt by intelligent orange growers."

The washes recommended for the other species might also be tried.

DESCRIPTIVE.

EGGS.—From 18 to 30 under each scale, less than .01 of an inch in length, ovoid, smooth, not quite as long as broad, of a bright yellow, promiscuously enclosed in body walls of dead female.

LARVA, OR FEMALE.—Length of body less than .01 of an inch, nearly twice as long as wide, bright yellow, ovoid, much wider towards head, being the widest at thoracic segments; two very short anal setae, hinder margin rough from numerous small fleshy tubercles, with a few short hairs around margin, no indentation like C. RUSCI, &c. ANTENNÆ, six-jointed, (not easily made out with my microscope, which is of a low power,) basil joint short and stout, nearly as wide as long, joints two and three smaller in width and of equal size, joints four and five about equal, longer and thicker than two and three, joint six much thinner, ending at tip into two long hairs, inner being longest, an inner and outer hair on basil joint, with two inner hairs and two outer above these; LEGS, with single jointed tarsi, ending in feeble claw and four digituli, the two upper being longest, femora thickly swollen, with a distinct lobe near base, from which a sharp spine issues, (Plate 4, fig. 9.)

SCALE.—Form round or circular, flattened, slightly rising towards centre, of from a reddish to blackish brown color, paler at margin, measuring from .04 to .12 of an inch in diameter; in the centre is a slight circular depression, in large specimens .02 to .03 of an inch in diameter, and of a bright golden yellow, with a small brown cap.

PLATE IV.

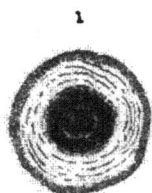

EXPLANATION TO PLATE IV.

Figure 1.—Scale of *Chrysomphalus ficus*, enlarged.

Fig. 2.—Scale, showing larval form beneath.

Fig. 3.—Immature larval form.

Fig. 5.—Fully matured larval form.   A depression caused by golden centre in top of scale. ·

Fig. 4 and 6.—Insect soon after hatching.

Fig. 7.—Egg.

Fig. 8.—Form of Egg-shell just after the insect has hatched.

Fig. 9.—Hind leg, showing lobe with hair.

Fig. 10.—Right antenna.

THE OVAL SCALE.

(*Aspidiotus citricola*, Packard.)

[Ord., HEMIPTERA. Fam., COCCIDÆ.]

BIBLIOGRAPHICAL.

Aspidiotus citricola, Packard, Guide to Study of Insects.

Prof. Packard, in his Guide to the Study of Insects, page 168, states this insect as having been found on the orange trees in Florida by Townend Glover, and gives it the name of *Citricola*, stating at the same time the possibility of its proving identical with Boisduval's *A.citri*, which was found damaging the orange crops in the Maritime Alps of Northern Italy. It. belongs properly to the genus Aspidiotus. In my opinion, it is probably identical with Signoret's *limonii;* but from my present knowledge of the Coccidæ, I do not feel warranted in positively stating it to be so.

ITS IMPORTATION AND SPREAD.

The only account of its introduction is that given by Glover in the Agricultural Report for the year 1855. He says : "While on the subject of Orange Scale Insects, it may be well to mention that some time last year (1855) another *coccus* was imported into Jacksonville, Florida, on some lemons sent from Bermuda; and as they may, perhaps, spread in the vicinity, it would be well to draw attention to the insect, and describe it as far as known. The length of the fully grown female scale is rather more than the twentieth of an inch ; it is somewhat pear-shaped, and of a brown color ; the grub is of a reddish-yellow, and furnished with a piercer from its breast, like the coccus first described ; the young have two antennæ, six legs and two long hairs or bristles, at the end of the body. The male scale is not so large as the female, and is formed of a white cottony or parchment-like substance, constituting a case, with an elevated and rounded ridge in the centre, in which a reddish pupa was found. The mouth of this case was stopped with a dark looking substance, apparently the cast skin of the larva. The male larva is reddish in color, and measures not more than the fortieth of an inch in length. The perfect fly is also red, and is furnished with two hairy antennæ, six legs, and has the thorax very large. The two wings are transparent, and the end of the body is furnished with a curved hard projection. As it is very probable that this insect will increase, it would be well to note any progress it may make during the ensuing year, and to use the remedies suggested in the first article on the coccus on the orange."

This well timed warning was not heeded, and the consequences followed.

ITS DISTRIBUTION.

It is now pretty widely distributed through Florida, being carried from place to place by the winds, on fruit, &c., but is still found much more abundantly in and surrounding Jacksonville. It is also found in Louisiana, the West Indies, Southern Europe, Australia, and probably wherever the orange is cultivated.

ITS NATURAL HISTORY.

Unlike the long scale, which is generally found on the twigs and branches, these species seem to be confined to the fruit and leaves, disfiguring the former to such an extent as to damage the sale. In an injurious point of view, it does not compare with the others; still, it injures the appearance of the fruit, and every possible means should be employed to prevent its increase and spread.

The ovoid or pyriform scale, sometimes nearly round, is of a light yellowish brown, and averages from .03 to .05 of an inch in length, with a very thin lighter margin. Under it, the pupa-formed female (Plate 1, fig. 9) is encased in a white cottony substance. It is purplish, with numerous small fleshy tubercles surrounding hinder edge, with a yellow anus.

The eggs (Plate 1, fig. 11) are of a pearly white color, and less than .01 of an inch in length, and somewhat pointed at one end. They number from eighteen to twenty-five under each scale. The young are broadly oval, .01 of an inch long, somewhat resembling the long scale, excepting they have two caudal filaments and are of a purplish color.

ITS NATURAL ENEMIES.

Besides the Orange Aphelinus, Twice-Stabbed Lady Bug, and its larva, the Minute Scymnus and the Large Scymnus, there are three other insects that prey upon it—a Mite, the Blood Red Lady Bug, and a hymenopterous parasite belonging to the Chalcididæ family. These are all of the greatest importance in the destruction of this scale. The Mite may be known as "Glover's, or the Yellow Orange Mite," the Lady Bug as above, and the hymenopterous fly as "the Blue Yellow-Cloaked Chalcid" of the orange. All the above-mentioned enemies, excepting the last three, have been described under the Long Scale; we have, therefore, but the habits of these three to treat of and describe.

GLOVER'S OR THE YELLOW ORANGE MITE.

(Tyroglyphus Gloverii, Ashmead.

[Ord, ARICHNOIDEA. Fam., TYROGLYPHIDÆ.]

BIBLIOGRAPHICAL.

T. Glover, U. S. Agricultural Report for 1855.

Acarus? Gloverii, Ashmead, The Florida Agriculturist, Vol. 2, No. 67, 1879.

Acarus? Gloverii, Ashmead, Canadian Entomologist, Vol. XI., 1879.

FIRST NOTICE OF THE MITE.

In the U. S. Agricultural Report for 1855, I find the following account of this mite by Townend Glover.

"There are also found on the orange trees numbers of small mites which have frequently been mistaken for the young *cocci,* but they may be very easily distinguished by their activity from the young scale insects, which crawl about very slowly. The mites have eight hairy legs, somewhat like those of minute spiders, and are mostly of a yellowish color, although some are also found of a delicate pink hue. They are generally seen running among the stationary *cocci,* and may often be found concealed under the old scales; but whether they do any harm to the tree, or merely feed upon the dead or dying *cocci,* has not yet been satisfactorily ascertained."

Although the first discoverer, Mr. Glover did not name or characterize this mite, I therefore described and named it in his honor, *Gloverii*.

ITS LIFE HISTORY.

During the winter and spring of 1879–80, finding these mites very numerous, I studied up their life history, (on orange trees in my yard,) an account of which was published in the American Entomologist for April, 1880, and is as follows:

The mite belongs to the genus *Tyroglyphus*. The eggs, between two and three hundred, are laid in December on the under part of an orange leaf, generally close to a midrib or a primary vein, and frequently alongside of a scale. They are elliptical, of a reddish-yellow color, nearly twice as long as broad, and very finely granulated. Length about 1-500ths of an inch. From the middle of January until the middle of March, there hatch a six-legged mite, of a bright blood-red, with three or four oval black spots on hinder part of abdomen, and sparsely covered with long hairs, six of these, (two anterior, two posterior, and two lateral,) are much longer than the others. In from three to four weeks, these curl up their legs and form a sort of pupa, which, in a few days, changes into an eight-legged mite, having nearly the shape as before, only larger, broader and more flattened; with two short hairs protruding from the head, and of a lighter shade of red. In these stages they are gregarious, all living huddled together close to midrib. The eight-legged mite again changes its skin and becomes the active little mite described in Canadian Entomologist. Scales on the same leaf with these mites are always empty, proving they are beneficial to the orange grower.

THE BLUE YELLOW-CLOAKED CHALCID.

(*Signiphora flavopalliatus*, N. Sp.)

[Ord., HYMENOPTERA. Fam., CHALCIDIDÆ.]

This is a very anomalous chalcid fly, discovered by me in September, running over the leaves of orange trees infested with the oval scale.

ITS NATURAL HISTORY.

Its habits are similar to the orange aphelinus. I have watched several through my pocket lens, as they are not at all timid. They would run up to a scale, tap it with their antennæ, and if not satisfied with their inspection, would run off to another, and so on until they were suited, then backing around they seemed to insert their ovipositor, probably at the same time depositing an egg into the scale. The fly is a beautiful little creature, less than .02 of an inch long, robust, with head wider than thorax, three ocelli, three-jointed antennæ, first joint being long, second small and round, third long and wide, club-shaped; the abdomen is somewhat sharply pointed, with a rather long ovipositor in the end; the head and abdomen are bluish-black, while the thorax is orange-yellow; the wings are clear, iridescent, and strongly fringed or ciliated with long hairs, with shorter ones on their surface; the legs are pale yellow, and the hinder pair is furnished with an anomalous five-lobed appendage, where usually is the tibial spur. Plate 2, fig. 2 gives an excellent idea of our little friend.

Since writing the above, I have raised specimens from scales put in a glass tumbler, and find that it is parasitic on this species. Owing to the anomalus character of this fly I can find no genus to which it belongs. I therefore, propose a new one, under the name of *Signiphora*, (the token bearer.)

SIGNIPHORA NOV. GEN.

Form robust, polished, or shining; head much wider than thorax, three ocelli, triangularly arranged, labial palpi three-jointed; antennæ inserted in front between the eyes, rather close together, three-jointed; first joint or scape long, second small and round, third large and fusiform, (Plate 2, fig. 3;) thorax broad, not quite as long as abdomen; legs setaceous, with five-jointed tarsi, first joint longest; *hind tibia in place of the usual spine, furnished with an anomalous five-lobed appendage,* (Plate 2, fig. 15.) In this respect, differing from any known chalcid. Abdomen somewhat sharply pointed and ending in rather a long ovipositor, (Plate 2, fig. 5.) Wings well rounded and strongly ciliated, (Plate 2, figs. 6 and 8.) Coxæ almost touching.

SIGNIPHORA FLAVO-PALLIATUS, N. SP.—Female.—Length .02 of an inch, Robust, polished; HEAD bluish-black, much wider than thorax, three ocelli, black, two raised curved lines, one on each side of antennæ, EYES prominent, numerous facets; ANTENNÆ three jointed, first joint shorter than third, wider and rounded at apex, second joint very small and round; apical, or third joint, longer than first, six or seven times longer than second, and widening very much, claviform; THORAX stout, nearly as wide as long, and of an orange-yellow, excepting a crescent shaped space (collare) next to the head, which is bluish-black; ABDOMEN longer than thorax, bluish-black, and decreasing sharply to a point, ending in rather a long ovipositor; UNDER SURFACE uniform bluish-black, with a few hairs on the different segments; WINGS hyaline, iridescent and strongly ciliated, well rounded at apex, with short setæ on the surface; LEGS pale yellow, with five jointed tarsi, setaceous, femora, somewhat swoolen. Instead of a tibial spur on hinder legs, there is a singular anomalous apical five-lobed appendage, (See Plate 2, fig. 15,) also two exterior spiny processes—COXÆ not quite touching each other. Male not yet discovered. Inhabits Florida. Described from numerous specimens.

THE BLOOD-RED LADY BUG.

(*Cycloneda sanguinea*, Linn.)

[Ord., COLEOPTERA. Fam., COCCINELLIDÆ.]

This well known species is widely distributed over the United States. In Florida it is particularly numerous on oak shrubs, evidently attracted there by a species of aphis.

Fig. 9.

ITS NATURAL HISTORY.

The larva of this species is entirely different from that of the Twice-Stabbed Lady Bug, being devoid of spines, flattened, with transverse yellow bands and spotted with black. It is most abundant in the spring, and is exceedingly active and voracious, running about in search of scale insects or aphides; on seizing one, it stops and immediately begins to devour it, then starts off in search of others, seemingly unable to appease its appetite. On reaching maturity, it fastens itself to a leaf by secreting a gummy substance, and gradually transforms into a naked pupa, changing in a few days to a perfect beetle, which is red, without spots or markings of any kind. (Fig. 9.)

THE BROAD SCALE.

(*Lecanium hesperidum*, Linn.)

[Ord., HEMIPTERA.   Fam., COCCIDÆ.]

BIBLIOGRAPHICAL.

Coccus hesperidum, , Linn, Sys. Nat., 1735.   Id. Faun. Suec., 1746.   La Hire et Sedilean, 1692.   *Coccus hesperidum*, Hist. Acad. Sciences et Mem. Acad. Sciences, 1704.   Reaumur Frist. Ins., 1736.   Geoffroy Ins., 1762.   Sulzer Ins., 1861.   Schaeffer Element, 1766.   Modeer Act. Gothend, 1778.   Gmelin, Sys. Nat., 1788.   DeVillier's Sys. Nat., 1789.   Olivier, Encyclo. Meth., 1791.   Fabricius Ent. Syst., 1794, et Syst. Ryng. Schrank Emon. Anst., 1783.   Fonscolombe, Ann. Soc. Ent., 1834.   Burm, Handb. Ent., 1835.   Blanchard, Hist. Nat. Ins., 1840.   *Calymnatus hesperidum*, Costa Nuev. Observa, 1835. Costa Faun. Ins. Nap. Gallinsects, 1837.   Lubbock, Proc. Roy. Soc., IX, 1848, and Ann. Nat. Hist, 1839.   Beck, Trans. Micros. Soc., London, new series, 1861.   Boisduval Ent. Hortic., 1867. Targioni, Catal., 1868.   (Signoret.)

This is another scale insect found on the leaves of the orange, and is by far less numerous and less to be feared than any of the others.

ITS IMPORTATION AND SPREAD.

Like the Long Scale, the Mealy Bug, the Oval Scale and the Red Scale, it has been imported, but in what year or about what time, it is impossible to find out.

ITS DISTRIBUTION.

It and *coccus citri*, are the only two scales mentioned as injurious to orange trees in Risso and Poiteau's elaborate work on L'histoire des Oranges," both of which are quite widely spread throughout Southern Europe.   Although known for many years to infest the orange trees of Florida, it has confined its attacks chiefly to the leaves, and has not spread rapidly, nor done much mischief.   It is seldom seen, except early in the spring and sometimes in the fall.   Why it has not increased more rapidly, is remarkable and unaccountable.

Fig. 10.

ITS NATURAL HISTORY.

This scale (Fig. 10, after Glover, Plate 1, fig. 12) is one of the largest found on the orange. It is oval, somewhat elongated, averaging from .08 to .14 of an inch long, and but slightly wider at one end. It is of a greenish brown color, highly convex, with wide flat margin surrounding the convex part, a posterior indentation with two lateral ones on each side, caused by the anus and legs. (See Plate 1, fig. 10.) The convex part is also punctured with large, irregular red indentations, which, as the insect reaches maturity, disappear, and the scale becomes dark brownish. The larva is elongated, semi-transparent, (See Plate 1, fig. 10,) with the viscera distinctly seen under the scale. The young, when first hatched, (Fig. 10,) are a little over .01 of an inch in length. They are yellowish, with two long anal filaments, resembling very much those previously described. They may, however, be readily distinguished from them by their having seven-jointed antennæ instead of six.

ON THE DIGESTIVE AND NERVOUS SYSTEM.

Sir John Lubbock, in Vol. XI of the Microscopical Society of London, gives a very interesting account of the digestive and nervous system of this insect. He shows that there is a great deal of variation in the intestines of different specimens. (Plate 1, figs. 13 and 14, represents the usual form.)

Plate 1, fig. 14.—G, G, hepatic glands. A, œsophagus, long and narrow, or stomach. F, pyriform crop bag or stomach, with a remarkable cellular contorted internal gland. D, ilium, short intestine opening into the rectum, C. B, narrow tube leading into rectum, which opens into vent on upper side of body. H, H, recurrent intestines, two ends of which are attached to the stomach, F. E, cœcum, swoolen at its base, and probably the equivalent of the sucking stomach.

THE YELLOW CHALCID OF THE ORANGE.

(*Trichogramma flavus*, N. Sp.)

[Ord., HYMENOPTERA. Fam., CHALCIDIDÆ.]

This little chalcid was first detected upon the leaves of some orange trees early in September. Its habits are, no doubt, similar to the others. I cannot state positively to which scale its attacks are confined, but as I found one under *Lecanium hesperidum*, I presume it preys upon it, and may account for the scarcity of this scale.

DESCRIPTIVE.

TRICHOGRAMMA FLAVUS, N. Sp.—Female.—Length .04 of an inch. Head wider than thorax, brownish, three ocelli triangularly arranged, with two smaller red ones back of these. Eyes reddish, excepting dark spot on side nearest ocelli. Antennæ, five-jointed, yellowish red, first joint longer than two and three combined and narrower than joint two. Joint two not as long as first, but wider—as long as joints three and four together. Joints three and four equal, narrower than second. Fifth, or apical joint, as long as second, third and fourth combined and much wider, claviform. Thorax and abdomen a bright yellow, reddish along hinder part of thorax, where it joins abdomen. Abdomen brownish on segments one to five around the spiracles, also a few hairs issuing therefrom. Ovipositor long, surrounded at base with short hairs. Wings hyaline, fore wings rather long and well rounded, with fringing of short fine ciliæ. Hind wings narrow, curving into a sharp point from the middle, also ciliated. Under surface uniform yellow. Legs thin, paler but uniform in color, sparsely covered with hair, a short tibial spur, tarsi five jointed. (Plate 1, Fig. 4.) Described from four specimens.

THE MEALY BUG.

(*Dactylopius adonidum*, Linn.)

[Ord, HEMIPTERA. Fam., COCCIDÆ].

BIBLIOGRAPHICAL.

Coccus adonidum, Linn., Syst. Nat., 1767. *C. adonidum*, Geoff. Ins., 1764. Fab. Syst. Ent., 1775. Id. Spec. Ins., 1781. Id. Syst. Ryng., 1801. Id. Ent. Syst., 1794. Modeer Act. Goth., 1778. Gmelin, 1788. DeVilliers, 1791. Foureroy Ent. Paris, 1785. Oliv. Encyc. Meth., 1791. Haworth, Ent. Trans., 1812. *Diaprostetus adonidum*, Costa, 1828. *C. adonidum*, Bouche Gart. Ins., 1833. Burmeister, Handb. der Ent., 1835. Blanchard, Hist. Nat. Ins., 1840. Id., 1843. Dict. Univ. Hist. Nat. article Cochenille. Amyot et Serville Hist. Nat. Ins. Hem., 1843. *Trechocorys adonidum,* Custer Ruric, Gard. Chron., 1843. *Coccus Zumiæ*, Lucas, Ann. Soc. Ent. Fr., 1855. *C. adonidum*, Mitner Ins. nuisibles aux plants de Cafe a Ceylon, 1861. Boisduval Ent. Hist., 1867. *Dactylopius adonidum*, Turgioni Cat., 1868.

This insect, as the above bibliographical account shows, has been known to the entomologist for over a century.

ITS IMPORTATION AND SPREAD.

Having been imported on various hot-house plants, it has spread so rapidly as to be found abundantly on different plants, shrubs and trees in nearly every part of the United States.

ITS FOOD PLANTS AND INCREASE.

In Europe, it has long been known to infest certain plants and shrubs, and is particularly destructive to the Pine-apple.

In Florida, besides the Pine apple, it has lately attacked the orange, guava and grapevine.

So numerous has it become and so difficult to exterminate, that there are now very few orange trees, with the exception of groves in South Florida and the interior, which have escaped its ravages. If its progress is not soon checked, it will ultimately spread all over the State, and prove to be among the worst of the many injurious insects found on the orange.

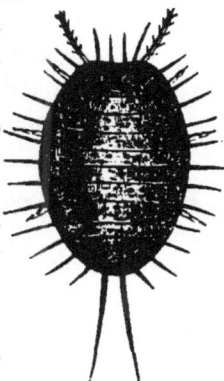

Fig. 11.   After Packard.

ITS NATURAL ENEMIES.

In the beginning of my studies, I unfortunately took it to be a new species, and gave it the name of Leaf Scale Coccus, *L. phyllococcus*, an account of which was published in the Canadian Entomologist and Florida Agriculturist for 1878. Subsequent researches have proved my error.

This particular species now under consideration, I find to be of an anomalous character among the *Coccidæ*. Unlike the others, it forms no scale for the protection of its eggs, and is not stationary, having power to move wherever it pleases. The eggs, instead of being laid under a scale, are placed beneath a cottony-like substance, secreted by the female. They are of a pale yellow, elliptical in shape, and about .02 of an inch in length. In from twelve to sixteen days these hatch, and the young are then of a yellowish color, .02 of an inch in length, oval, with two antennæ, six-jointed, the last joint being the longest and covered with a few short hairs; it has also two short anal setæ. (Fig. 10, after Packard,) gives a very good idea of it after the mealy substance has been removed. Surrounding the outer edge, are also short hairs. The young soon begin to run about, sucking or feeding on the tender leaves and shoots, and the fine mealy substance begins to be secreted from pores all over the body, hence the name Mealy Bug. When fully grown, this insect is .14 of an inch in length, having a round, spherical or globular form, and lays its eggs as described above. The mealy substance often becomes so thick as to form scales, which are attached to the hairs surrounding the body; particularly does it accumulate above the spiracles or breathing holes.

THE MALE.

For the past two years I have failed to detect the male of this species. This fall, however, I have been more fortunate, having succeeded in finding him caught in a spider's web on an orange tree badly infested with the Mealy Bug. Being nearly twice as large as the males of the other scale insects treated of in this book, it is therefore surprising that it should have remained so long undiscovered. It is brownish in color, and instead of a long caudal appendage like the male of *A. Gloverii*, he has two long filaments issuing from the sixth segment of the abdomen, and his head is separated from the thorax by a well defined neck.

DESCRIPTIVE.

Male.—D. ADONIDUM.—Length not quite .04 of an inch. Ala Expanse .08 of an inch. Brownish. Head separated from thorax. Eyes black, prominent, no facets discernible. No beak—in place are two large black smooth ocelli. Antennæ ten-jointed, first joint thick and stout, second as thick as first and rounding at top, third longest, much narrower than first and second, four five and six, about equal, seventh greatly swollen, eighth about same length but much thinner, ninth slightly longer, tenth greatly swollen, thorax slightly longer and much wider than abdomen, wider anteriorly than posteriorly, rounded in front. Wings hyaline, spatulate, three veins, costal parallel with outer edge, thickening at quarter and at apical margin, second vein starting at quarter of wing and crossing diagonally to hinder edge, the third has a small veinlet just beneath it. Abdomen pale yellowish, eight segments discernible, sixth widest and with long filament appendages springing out, one from each side, longer than abdomen, seventh small—blackish beneath, ninth almost a knob. Legs pale yellowish, semi-transparent, with very long narrow tibia, hinder tarsal joint swollen much larger than the others.

ITS NATURAL ENEMIES AND REMEDY.

No enemies have been detected preying upon this species, although three or four times I caught a large black ichneumon fly in close proximity to a cluster of them.. They increase very rapidly, breeding all through the year, and severe methods should be used for their destruction. The usual methods seem to have no effect. I would, therefore, recommend kerosene, diluted with three parts of water. This should be syringed over them. Great care should be taken to shake the wash well before applying it, for unless this is done, the oil rises to the top of the water, and wherever pure kerosene is ejected upon the tree, the leaves and twigs are sure to die. This is obviated to a certain extent by having the wash properly mixed; for then what few leaves do die are soon replaced and the trees left free from bugs.

THE ORANGE PSOCUS.

(*Psocus citricola*, Ashmead.)

[Ord., NEUROPTERA.   Fam., PSOCIDÆ.]

BIBLIOGRAPHICAL.

Psocus citricola, Ashmead, Canadian Entomologist, Vol. XI, 1879. The Florida Agriculturist, Vol. II, 1879.

ITS NATURAL HISTORY.

These active little insects are found very plentifully on orange trees badly affected with the Mealy Bug. At first, I thought they might possibly be beneficial by preying on the eggs of the latter, but careful observation fails to show any such disposition on their part. They are, however, not injurious, only feeding upon the excrementitious particles caused by the Mealy Bug. Their eggs, of a pearly pinkish white, are oval, about .02 of an inch in length and are laid in oval masses, from seven to ten, on the under, and sometimes on the upper part of a leaf, protected by a closely woven web, through which are sprinkled sooty particles.

The young, when first hatched, are exceedingly active, aphis-like looking creatures, and are fond of clustering together under the web. Here you can find them of all descriptions, from the small wingless ones to those fully developed. On disturbing them, they disperse with surprising rapidity.

DESCRIPTIVE.

P. CITRICOLA.—Elongate, pale yellowish. Head large, as wide as long, outer edge from eye to eye forming a perfect half circle. Eyes are large and very prominent. Maxillary palpi four-jointed, the basal joint little longer than either of the others, but narrower—the others about even in length but gradually increasing in thickness, the last being the thickest. Antennæ three-jointed, first two short, same size, as wide as long. The last joint is long and filiform, reaching nearly to the end of the abdomen and covered with long fine hair. Thorax narrower than head, slightly longer than wide, rounded at edges, with a transverse suture dividing it into two parts (immature specimen). Abdomen longer than head and thorax together, eight segments, the largest being nearly as wide as thorax. Legs six, rather long, tarsi two-jointed, ending in two minute claws. The abdomen and legs have small short hairs springing out all over them. Wings hyaline, with costal, subcostal, median and submedian veins. In fore wings the subcostal runs parallel with costal until before reaching apex it bends downwards and then curves upwards, ending at termination of costal vein, forming a stigmal cell which is opaque. It also sends a veinlet from before middle that descends and curves around upwards until near the third of the wing, when it divides, the lower ending in outer edge. The other runs to below apex, near the edge, where it divides into two short veinlets, terminating at outer edge. Medium vein curves slightly downwards until near the middle of wing, it then divides into two, the lower descending till near apex of inner edge, when it suddenly curves upwards, terminating at outer edge, the cell thus formed being opaque. The other veinlet ascends, crossing the branch of the subcostal till just before reaching the apex it breaks into two veinlets, forming a small triangular cell at apex. Hind wings contain one costal, three subcostal, two submedian, and one internal cell. Length of matured specimens from .10 to .12 of an inch.

THE LEAF FOOTED PLANT BUG.

(*Leptoglossus phyllopus*, Linn.)

[Ord., HEMIPTERA.   Sub-Fam., ANISOSCELIDINA.]

BIBLIOGRAPHICAL.

Cimex phyllopus, Syst. Nat. ed. 12, i. 731, No. 113.
Lygæus phyllopus, Fab. Ent. Syst., iv, 139.
Anisoscelis albicinctus, Say. Heteropt, New Harmony, 12, No. 2;
    Wolff, Icenes Cin., 196, fig. 190.
Anisoscelis confusa, Dallas, Brit. Mus. List Hemipt, ii, 453, No. 4.
Theognis phyllopus, Mayr, Novara Reise, Hemipt., 103.
Leptoglossus albicinctus, Stal, Hemip., Feb. 1, 52, No. 5.

Anisoscelis phyllopus, Barns, Handb., ii, 322, No. 5 ; Westw., in Hope Catal. ii, 16.

This is a curious shaped, reddish-brown bug, with a long sharp beak, and a transverse yellowish white band across its wing covers. These, when raised, show the back to be hollowed and flattened, of a bright red color, with transverse black spots. The shanks of the hind legs are flattened out into leaf-like appendages, (Fig. 12.)

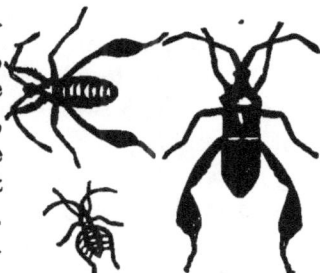

Fig. 12. After Glover.

At first, the young are of a bright yellowish red, without the flattened appendages to the hind legs. (See lower figure in cut.) These only appear just before casting the last skin.

ITS DESTRUCTIVENESS.

It has not only proved very destructive to the orange tree by sucking the sap from the tender shoots and terminal branches, thus killing them, but I have also observed it thrusting its sharp beak into plums and sucking their contents to such an extent as to render them unfit either for eating or selling. On Mr. T. W. Moore's place, at Fruit Cove, they were unusually abundant in the summer of 1879, and in August of the same year, I observed them in countless numbers, settling upon the heads of the young rice, just before ripening, from six to ten on a blade, in various stages of growth, sucking the " milky kernel." Mr. M. informed me that nearly all his rice crop was destroyed by them in this manner.

REMEDY.

The only method of destruction I can at present suggest is to catch them in a butterfly net and scald them.

THE TREE PLANT BUG.
(*Brochymena arborea*, Say.)
[Ord, HEMIPTERA. Sub-Fam., HULYDINA.]

BIBLIOGRAPHICAL.

Pentatoma arborea, Say, Journal Acad., Phil., iv, 311, No. 1.

Complete writing, ii, 239. Dallas, Brit. Mus. List, Hemip. i, 188, No. 1.

This is a large speckled grey plant bug, not so numerous as the other, and of a different shape. It is of a flattened, oval form, .60 to .64 of an inch in length, sides of thorax being armed with seven small spines; the antennæ thin and five-jointed, reaching to end of first ventral segment; the beak is long and slender, and when not in use lies folded close to the under part of the body, .reaching to second ventral segment. I have often caught it on the trunks of orange trees infested with scale insect, and think it probably feeds on them, as I have never detected it feeding upon the young shoots. It is therefore questionable whether it is beneficial or injurious.

THE RUST MITE OF THE ORANGE.

(*Phytoptus oleivorus*, Ashmead.)

[Ord., AOARINA.   Fam., PHYTOPTIDÆ.]

BIBLIOGRAPHICAL.

Typhlodromus oleivorus, Ashmead, Canadian Entomologist, Vol. XI, 1879. Florida Agriculturist, Vol. II, 1879.

Some of the four-legged mites that are now known under the family name of *Phytoptidæ*, or popularly Gall-Mites, have been known to entomologists for two centuries, and their life history, up to the present time, remains unravelled. The late Andrew Murray, in his work entitled "Aptera," gives a succinct account of all known species, and to those desiring fuller information on these interesting mites, I would recommend his work.

ITS DISCOVERY.

In the latter part of August, 1879, the Rev. T. W. Moore called my attention to this interesting mite, at the same time stating his belief that it was the cause of the orange rust. I immediately began to study it, and soon after wrote him that I had discovered what it was, and forwarded a description of it for publication, crediting him with the discovery, under the name of *Typhlodromus oleivorus*, i. e., oil-eating, from supposing it to feed on the essential oil of the orange. He replied as follows:

Mr. W. H. Ashmead:

Dear Sir:—Your favor of yesterday is at hand. You are mistaken in mentioning me as the discoverer of the insect. The opinion has long been entertained that the rust was occasioned by an insect, but we had no certain knowledge of the fact till a few months since. Mr. J. H. Gates, from near Palatka, discovered something, and placing it under the microscope, found it to be an insect. Mr. W. C. Hargrove was the first to mention the subject to me.

I am still at work studying the habits and looking for the destroyer of the insect, and think I am making some progress, though some of my experiments have been rather expensive.

There is one form of rust which I think is not caused by this insect, as evidently the oil flows out without the assistance of the puncture of the insect, simply from its superabundance.

I am, my dear sir, yours truly, T. W. Moore.
Fruit Cove, August 27.

The article was as follows:

My attention has been drawn by the Rev. T. W. Moore to a strange insect on his orange trees and on the trees of Mr. Byron Oak, and I have also noticed it on the fruit of several others in and surrounding Jacksonville. It is one of the most interesting insects I ever saw. To-day, August 25th, I obtained specimens off Mr. Oak's trees, and have examined them thoroughly with a powerful microscope, of six or seven hundred diameter. They are not apparent to the naked eye. I found them in great numbers on all rusty fruit examined; they are almost stationary, seldom moving, and attach themselves to the oil cells, which they puncture with their beaks, probably, feeding on the oil. Mr. Moore attributes to them the cause of the orange rust, in which opinion I concur.

The puncture causes the oil to exude, the chemical action of the atmosphere causes it to oxidize, and the result is a hard, rusty skin. On all oranges that had begun to rust, we found the insects in great numbers; nor could we find them anywhere else, even after a careful examination. Half an hour after the fruit

had been picked, no insect could be found; they had all fallen off
and disappeared.   This is the reason why the microscopist could
never detect any insect, and as a dernier resort attributed the
result to a fungoid.

It is a four-legged mite, belonging to the genus Typhlodromus,
and is probably the first species of the genus discovered in America.   It may be termed the oil-eating mite of the orange.

Thus the long vexed question of what causes the orange rust
is solved, and proves not to be a fungoid, as many supposed, but
an infinitesimal creature, that could never have been discovered
except with the aid of a microscope.

In the Florida Agriculturist, the following communication
appeared:

I see that Mr. Ashmead has classified and named the small
insect that I believe to be the cause of the rust on oranges.   I
will give a short history of its discovery.

Some time in May, Mr. J. H. Gates called my attention to a
small insect upon an orange he was just sending the microscopist
of the Agricultural Department at Washington.   I felt this was
a valuable discovery, and that the insect was the cause of the rust,
and at once commenced investigating the little "chap."   Mr.
Gates' reply from Washington did not, in my opinion, describe
the insect as I saw it with my glass.   I immediately sent specimens of the insect and oranges, together with a drawing of the
insect, to Mr. Thomas Taylor, Washington.   Unfortunately, his
reply was misdirected and did not reach me.   In a subsequent
letter asking that I send some leaves from the same tree, he says:
" I think that the insect you describe is the cause of the orange
disease in this case."   This much for the discovery of the insect,
which is due to Mr. Gates.

Early in June, Rev. Mr. Moore, of Fruit Cove, was in Palatka,
and I gave him a description of the insect; it being night, I could
not show him a specimen.   A few weeks after, Mr. Hargrove
showed Mr. Moore the insects in his own grove, and he has devo-
ted much time and attention to them since.

In my investigation I was aided by Mr. W. C. Hargrove, who
is a close observer, and has been much interested in the cause of
rusty oranges, his oranges having been rusty for some years.

He said at once, as soon as I described the appearance of the oranges, that each year his oranges had that appearance at the commencement of the season. I considered this good testimony and valuable information.

At first the orange had a dusty appearance to the unaided eye, and although they are mere "mites," after a little practice you can detect them at once. Under a glass, the orange appears covered with thousands. I have examined oranges from different groves, and in every case have found the insect where there were rusty oranges. They remain upon the orange about four weeks—that is, when there is a large number upon one orange. By that time they have punctured all the oil cells, which allows the oil to exude, and with the excretion from the insect become oxidized, and they, no longer able to obtain food, leave. I find the oranges attacked early in the season are more uniformly rusty and darker than those later in the season, which may be owing to two causes—not so many insects, or the oranges more fully developed.

They prefer the oranges on the outside of the tree and most exposed to be seen. I have asked: How do you account for so many turning dark immediately after a rain? I account for it in this way: The insect has been doing its work, and the rain coming in contact with the oil and resinous matter, causes it to be precipitated at once, which hardens, and as soon as the sun shines out bright, turns dark by oxidization. By taking an orange, before it had turned dark, but had that peculiar dust-like appearance and plenty of insects, dipping it into water, the resinous matter was precipitated at once, and could be removed by the thumb-nail quite readily, if done before it had time to dry and harden to the orange.

About three weeks ago they made their appearance on an orange tree that had fine bright fruit. To-day over half of the fruit is of a bronze color, not very dark, but so much so that their price will be materially lessened.

PALATKA, FLA., Sept. 11.   E. S. CRILL.

Credit is therefore due Mr. J. H. Gates as the original discoverer, which Mr. Crill says was made the last of May, 1879. Such an important discovery as this should have been published at once.

WHAT IS THE ORANGE RUST ?

For the past ten or fifteen years there has appeared on vast numbers of the fruit, a reddish or brownish rust, which was variously attributed to a fungus, moisture, wet swampy soil, proximity to pines, or to an insect. All these theories were promulgated from time to time, but no facts given to support them. I consider this question now definitely settled beyond controversy as the work of the *Phytoptus.*

ITS ORIGIN AND SPREAD.

As far as I can learn, the rust first made its appearance soon after the war on fruit from. Mandarin, and it has since spread nearly over the whole of Florida, some years appearing more plentiful in one place than in another, and without regard to soil, swampy land, proximity to pine trees, or anything else.

DAMAGE DONE TO THE FRUIT.

The reddish or brownish rust does not materially injure the fruit or tree, excepting in appearance, which damages the sale. In fact, for shipping purposes, it is rather beneficial, as the hard rusty skin prevents the air from reaching the interior part of the fruit, thus causing it to keep much better and longer.

ITS NATURAL HISTORY.

As I have previously stated, nothing, so far, is known of the life-history of any of these four-legged mites. While

"The garden glows and fills the liberal air
With lavish fragrance; while the promised fruit
Lies yet a little embryo unperceived
Within its crimson folds,"

the little *phytoptus* begins its work of destruction.

Last April, I noticed numerous gall-like *prominences* on the orange leaves, and it immediately occurred to me that these might probably prove to be the work of the little *phytopti*. I began investigating by placing some under the microscope, and dissecting others. 1 found that they were caused by an insect, but whether by the phytoptus or not I cannot positively say, as' I left Jacksonville the last of April, and could not continue my researches. My opinion is that they were the cause. It is probable that they hatch about the time the fruit begins to form, and immediately crawl from the leaves to the fruit. Plate 2 and plate 4, fig. 11, highly magnified, will give one a good idea of the little fellow.

It is probable that "rusty fruit" could be prevented by an early application of one of the washes used for scale insects. Mr. Moore informs me he has had no difficulty in killing them with a wash made from tobacco leaves and soap suds.

PHYTOPTUS OLEIVORUS. Whitish flesh color, elongated, cylindrical, gradually increasing in size near the head it becomes twice as thick as posteriorly; abdomen finely and transversely striate, apparently consisting of numerous very thin segments; at the extremity is a biped appendage that evidently assists in clinging to the orange; just above it protrude two caudal filaments; head almost hidden in thorax; four legs rather short with one claw, a long hair springing from knee.

THE ORANGE BUTTERFLY AND ORANGE DOG.

(*Papilio cresphontes*, Cramer.)

[Ord., LEPIDOPTERA. Fam., PAPILIONIDÆ.]

BIBLIOGRAPHICAL.

Papilio thoas, Fab., Spec. Ins. Id. Mantis Ins. Id. Entom. Syst. *Papilio cresphontes*, Hubst. *Papilio thoas*, God. Enc. Meth. Ins. *Papilio thoas*, Boisduval et Leconte, Lepidopteres de l'Amerique Septentrionale, 1833. *Papilio thoas*, Boisduval, Species General des Lepidopteres, 1836. *Papilio thoas*, Synopsis of described Lepidoptera of North America, by Morris. *Papilio cresphontes*, Cramer, Die Uittandsche Kapelten Voorkomende in de drie, Waereld Dulen, Asia, Africa en America, Plate CLXVI, fig. B.. 1779.

One of the most beautiful and abundant butterflies to be seen in Florida from early spring until late in winter, is a large black butterfly with two yellow bands extending across the wings, formed by a series of large yellow spots, (Fig. 13.) As early as January and February, it may sometimes be seen flying up the principal business thoroughfare of Jacksonville; now in the middle of the street, now on the sidewalk, bobbing here and there over the heads of the people, then off over the highest stores, into some neighboring yard to sup the sweets from flowers.

Few are aware that this gay, active butterfly is the parent of
that ugly, curious excrementitious looking caterpillar, so well
known to orange growers, under the common name of Orange
Dog.

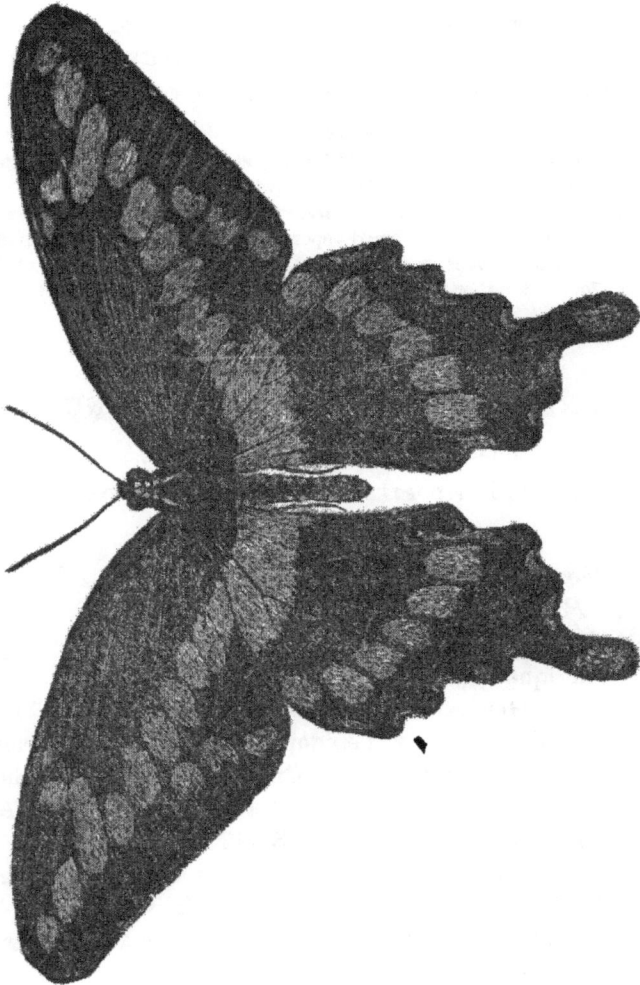

Fig. 13.

ITS DISTRIBUTION.

This butterfly was first described by a Dutch Entomologist,
Cramer, one hundred and one years ago, (1779) under the name
of *Papilio cresphontes*. It has a wide geographical distribution,
being found in Brazil, the West Indies, most of the Southern

States, and extending Northward as far as Ontario and Canada, in which place it is rare.

ITS FOOD PLANT.

Its original food plant was probably the Prickly Ash, *Zanthoxylum Americanum*, or an allied species. It is here found feeding upon *X. Carolinianum* and all species of the *citrus* family. On the Florida Keys, I found its food plant to be *Zanthoxylum Pterota*, a small species of ash. In Kansas, Prof. Snow reports it feeding on Prickly Ash, *Zanthoxylum Americanum*, Hop Tree, *Ptelea trifoliata*. In Ontario, Canada, Wm. Murray, Canadian Entomologist, Vol. XI, states having found it feeding on *Dictamnus fraxinella-rubra*.

ITS NATURAL HISTORY.

The female butterfly, when ready to oviporate, flies here and there through the grove, depositing from one to two round eggs on a leaf. From

Fig. 14.

these, in a few days, hatch out a small hairy, brown colored caterpillar, without sign of the cream colored band over the back; this does not appear until the third moult. After this moult the hair disappears, and it has a slimy nauseating appearance. It feeds on the orange leaves, and after several moultings of its skin, reaches maturity, measuring from two to three inches in length. (Fig. 14.)

On disturbing, it protrudes two long red filaments from the segments above the head, and at the same time secreting a very disagreeable odor, which is supposed to be a protection against its enemies. When ready to pupate, it secretes a small silky web to a twig or the body of the tree, and attaches itself by its tail and a silken thread across its back, then changes into a chrysalis, (Fig. 15.)

In from eight to sixteen days, the back of the chrysalis splits and the butterfly emerges. At first, the wings are wet and crumbly, but the butterfly, crawling up the tree or twig, secures a firm hold with its feet, stops, raises its poor de-

Fig. 15.

formed looking wings, and as the vessels fill with air, they expand. It continues slowly flapping them up and down, until they are entirely expanded and strong, when it is ready for flight. The fall brood of caterpillars remain in the chrysalis all the winter, transforming into butterflies early in the spring.

REMEDY.

The caterpillar of this butterfly, unlike many others, has no parasite to prey upon it, or at least none so far discovered. Even birds, on account of its ugly, unpalatable appearance, will not eat it. The question is how to keep them in subjection and prevent them from defoliating the young orange trees, of which they are particularly fond. The only remedy I know of, is hand-picking, a tedious process, but a very necessary one, especially in the spring and fall, when they are most numerous. The *modus operandi* is to provide yourself with a bucket or something to keep the caterpillars in. Go carefully over your trees, pick off all to be found, put them in the bucket, then kill by scalding them. Repeat this every two or three weeks and your trees will be kept comparatively free from them.

THE LANCE RUSTIC MOTH.

(Agrotis Ypsilon, Hufn.)

[Ord., LEPIDOPTERA.   Fam., NOCTUIÆ.]

BIBLIOGRAPHICAL.

Agrotis Ypsilon, Hufn. Suffusa Den. and Scheit. Wiener Verzeichniss *Telifiera*, Harris, Insects Inj. to Vegetation.

In February, 1880, Dr. Robert S. Turner, of Fort George Island, Florida, brought me pupæ of a nocturnal moth, dug up from the roots of his orange trees. These, I put in a breeding box, and by the first week in March, there hatched out several moths with which I was unacquainted. I sent specimens to Professor Riley, who replied as follows:

"DEAR SIR: The moth you send is a dark specimen of Agrotis Ypsilon, Hufn. This moth is common to both Europe and America, and occurs even in New Zealand. It was subsequently described in *Wiener Verzeichniss* under the specific name of *suffusa*, and again by Harris, under the specific name of *telifera*, by which name I have given an account of it in my First Report on Insects of Missouri, p. 80. It may popularly be known as the Lance Rustic, and is the parent of one of our commonest 'cut worms,' illustrated on plate 1, fig. 9, of report cited. This worm does great injury by cutting off tender plants early in Spring. It comes to its growth in the latitude of St. Louis, during May, the moth making its appearance two or three months later. It is such a general feeder that I am not surprised at your finding it on the leaves of an orange tree."

ITS NATURAL HISTORY.

Professor Riley describes the larva under the name of the "Large Black Cut-Worm. It is an inch and a half in length when crawling, and its general color above is dull, dark brown, with a faint trace of yellowish white along the back. The subdorsal line is more distinct, and between it and the stigmatic are two other indistinct lines. There are eight black shining, piliferous spots on each segment; two near the subdorsal line, the smaller a little above it. The other two are placed each side of the stigmata, the one anteriorly a little above the other just behind in the same line with them, and having a white shade above it." (Fig. 16, upper figure.)

This caterpillar feeds only at night, crawling down from the orange tree in the morning and hiding in the ground or under rubbish during the day. The eggs of this species are of a flattened spheroid shape, purplish in color, with longitudinal ribs, and resembles very much the eggs of the cotton moth. They are laid in clusters on under or upper part of an

Fig. 16.

orange leaf, and are subjected to the attacks of a parasite or chalcid fly, which, as I have never bred, cannot describe. I have frequently found the eggs after the fly had escaped. These eggs appear to be laid in early spring, as this is the only time I have been able to procure or find them.

The moth (fig. 16) is nocturnal in its habits. The fore-wings are light brown, shaded with lines and bands of a dark brown; the hinder are pearly white, dusky around the edges with a narrow fringe of down.

REMEDY.

They could be considerably reduced in numbers by looking for and destroying the eggs in spring.

THE WOOLLY BEAR OR STINGING CATERPILLAR.
(Lagoa opercularis)
(Ord., LEPIDOPTERA. Fam., BOMBYCIDÆ.)

Towards the latter part of August, until cold weather sets in, one frequently comes across a curious looking caterpillar crawling on a rosebush, orange, or plum tree, about an inch in length and thickly covered with long yellowish brown hairs. Under these long hairs are others, sharp and barbed, which, should one happen to come in contact with, are capable of inflicting a very severe sting. An instance was related to me of a boy who was severely stung by one of these worms, and so powerful and poisonous was the wound, that it was thought to be the bite of a snake, until investigation proved it to be one of these small Woolly Caterpillars.

ITS NATURAL HISTORY.

On attaining full growth, this caterpillar makes a thin, tough, oval, dark brown cocoon, the inside of which is white and smooth, while on the outside are a few silken hairs. This is attached to under part of boards, logs, &c. In this it changes to a pupa, and in time transforms into a yellowish brown moth, which, on account of being covered with fine hairs, is termed the " Woolly Bear."

ITS FOOD PLANTS.

The caterpillar, in Florida, has been found feeding on the leaves of the orange, the quince, and other trees; also on the rosebush.

ITS ENEMIES.

We have raised from the caterpillar several specimens of a "Tachina Fly," not yet determined.

THE BASKET WORM OF THE ORANGE.

(*Platœceticus Gloverii*, Packard.)

[Ord., LEPIDOPTERA.   Fam., BOMBYCIDÆ.]

BIBLIOGRAPHICAL.

Packard, Guide to the Study of Insects, page 291.   Glover, U. S. Agricultural Report, 18.

ITS NATURAL HISTORY.

This curious bombycid case worm was named by Professor Packard after its discoverer, Townend Glover, who published a description of its habits in the Agricultural Report for 1858, and from which I shall quote.   He says : " A small hang or drop worm (*oiketicus*) is very prevalent upon orange leaves, and is found most frequently suspended from the leaf, entirely enveloped in a brownish, oblong, oval case, of a paper-like substance, spun by the worm within and interwoven with dried scraps of the leaf itself, or any other material over which the worm may wander."

The male case is about 0.5 of an inch in length.   Within it lives the worm, which is 0.3 of an inch in length, and of a brownish color, clouded or spotted with darker brown on the head and two first segments, protruding only its head and first segment when feeding or moving from place to place.   Within this case, which is spun from the worm itself, is found a chrysalis of a dark brown color, which, when about to change, pushes itself nearly out of the opening in the lower extremity, when the back splits and a small moth of about 0.5 of an inch in breadth across the expanded wings of black color and having a feather-like antennæ, comes out.

The female case is much larger, measuring 0.7 of an inch in length, and is formed of the same materials as the male case. The female, however, never acquires wings, but when ready to change, fastens the case to the leaf with silk, lays its eggs and dies in the case which it had constructed as a shelter for its soft and fleshy body.

The eggs are likewise laid in the case, and the young, when hatched, escape from the orifice at the lower end and disperse over the tree in search of food."

I have also found this insect feeding on the leaves of the fig, and have also detected it, or a similar species, preying upon the scale insects. It is by no means common, nor do I apprehend its ever becoming so. Should it be proven to be identical with the species detected preying on the scale insects, it must be considered beneficial. Fig. 17, (after Packard,) gives a good idea of its

Fig. 17.

different forms. A, moth; B, caterpillar; C, pupa; D, case in which caterpillar lives.

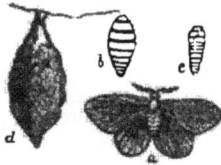

THE ANGULAR-WINGED KATY-DID.

(*Microcentrus retinervis*, Burm.)

[Ord., ORTHOPTERA. Fam., LOCUSTIDÆ.]

This large, green Katy-did, or Grasshopper, as it is variously called, is among the commonest we have. During the daytime it is seldom seen, being hidden away among the thick foliage of trees and shrubs; but towards dusk, as the shadows of night begin to fall, it comes forth from its hiding place and begins its song of "Katy-did and Katy-did'nt," so familiar to every one. This song is not produced by its mouth, as nearly every one supposes, but by slightly opening its wings and rubbing them against its thighs.

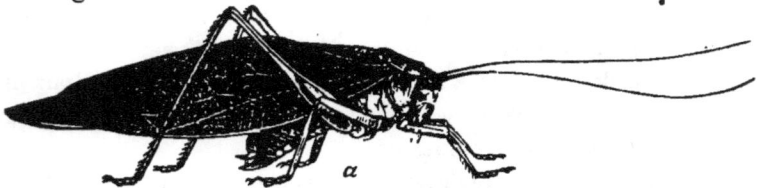

Fig. 18. (After Riley.)

ITS NATURAL HISTORY.

Early in the fall, during the winter and in the spring, one may often observe on the outer edge of the orange leaves two parallel rows of large mussel-shaped eggs of a greyish slate color. These are the eggs of the Katy-did. Prof. C. V. Riley, who has observed it closely and bred it in confinement, gives an excellent account of its method of ovipositing, in his Report on Missouri Insects, which I shall quote at length:

" The female commences to oviposit early in September, and continues to lay at intervals until the first severe frost. The eggs are occasionally deposited during the day, but the operation usually takes place at night. Selecting a twig of about the size of a common goose-quill, this provident mother prepares it for the reception of her eggs by biting and roughening the bark with her jaws for a distance of two or three inches. This bite is not gradual, like that made when feeding, but is sudden and vigorous, the insect chewing and pressing the twig each side, so as to form an edge. This operation is accomplished by a sudden nervous shake of the body from side to side, and lasts sometimes but two or three minutes, sometimes more than ten. When the operation is accomplished to her satisfaction, she clutches with her front feet the stem to be used, and anchors the middle and hindmost feet for the most part upon contiguous leaves or branches, and often quite wide apart. Then, if she has her head in an upward direction, (for it seems to be immaterial to her whether the eggs are placed from below, up, or *vice versa*,) she begins at the lower end of the roughened portion of the twig, and, after fitting it anew with her jaws and measuring and feeling it over again and again with her palpi, as if to assure herself that all is as it should be, she slowly—with much apparent effort, and not without letting it partly fall several times—curls the abdomen under until the lower edge of the curled ovipositor is brought between the jaws and palpi, by which it is grasped and guided to the right position.

"It is then worked slightly up and down for from four to six minutes—all the time guided by the jaws—while a shiny viscid fluid is given out apparently from the ovipositor. Finally, after a few seconds, rest or suspension of the work, the egg gradually rises, and, as it passes between the ovipositor, runs so that the one end appears almost simultaneously from between the convex edge with the other from the lower tip of the blades. The egg adheres to the roughened bark in an oblique position. It is first black and highly varnished, but it acquires its normal gray color within eight or ten hours.

"After the egg is placed, the abdomen is straightened out and the insect rests for a few moments, soon, however, to resume her efforts and repeat the like performance in every particular, ex-

cept that the second egg is placed on the opposite side of the
twig and a little above the first one. The third egg is pushed
in between the top of the first one and the twig, the fourth between the top of the second, and so on, one each side, alternately.

"Thus, these eggs are not laid, as we might naturally imply,
one over the other, but rather, one under the other; i. e., each
succeeding pair having their ends thrust in between the tops of
the preceding pair, the teeth at the end of the ovipositor helping
to crowd the end into place.

"The length of time required from the commencement of the
fitting of the twig to the proper placing of the egg varies all
the way from 5 to 20 minutes. Sometimes, as for instance,
where a bud comes in the way, the preparation of the twig will
require a comparatively long time, and after the ovipositor is
brought up and a futile attempt made to place the egg, it will be
let down again and the work of preparing the twig more vigor-
ously prosecuted the second time. The number of eggs laid at
one time varies from two to thirty, the first batches containing
more than those deposited later in the season. Each female produces from one hundred and fifty to two hundred, or perhaps
more, and I have known them to lay on the edge of a leaf, or of
a piano-cover, or along a piece of cord. These eggs, as already
remarked, are rather flat when laid, but become more swollen,
so that they have a narrower look as they approach the hatching
period in the spring.

" During the early part of May, the embryo larva, which lies
straight in its egg—completely filling it—with the legs bent up
as in a pupa, and the long antennæ curling around them, attains
its full development, and after hours of tedious contracting and
expanding movements, manages to burst the egg open at its top
or exposed end along the narrow edge, and generally about half-
way down. Through this opening young Katy slowly emerges,
undergoing a moult during the process, and leaving its skin in a
crumpled white mass attached to the empty bivalvular egg shell.
Including hind legs and antennæ, it measures at this time, rather
more than an inch in length, the body alone being one-eighth of
an inch long ; and in contemplating it, one cannot help but wonder how the long, stiff legs, and great length of the antennæ, together with the plump body, could so recently have been com-

pressed into the comparatively small skin to which we see it clinging.

"In from ten to twenty minutes after hatching, these little beings essay their first leaps, and soon begin to eat with avidity. They feed with almost equal relish upon a great variety of foliage, but I have found that when reared upon very succulent leaves, such as lettuce, cabbage, purslain and the like, they are less hardy, and do not attain so great an age as when nourished upon more ligneous food, as the leaves of oak, apple or cherry.

"The larval life of these insects lasts from seven to eight weeks. Shortly before the change to pupæ, which takes place towards the end of June, the rudiments of the wings and of the sexual organs may be distinguished. In the pupa state they are quite pretty, and their faces have a comically wild look, and every motion is invested with a sort of dignity that cannot fail to amuse the observer.

"Including the moult in leaving the egg, they cast their skins five times, becoming pupæ at the fourth and acquiring wings at the fifth. In each case the palpi are adroitly used to help the long antennæ out of the old skins, and a description of the last, which is more easily watched, will convey a correct idea of all. In changing from a pupa to the perfect form, the insect stations itself firmly upon a large stem or a couple of twigs, which branch in such a manner as to afford a convenient support, and, after a short period of inactivity, a rupture appears in the covering of the head and gradually extends backward to the posterior edge of the thorax. The armor of the head is next detached from the neck, and by a few upward and downward motions, is made to slide off in front, the long thread-like antennæ being drawn out of their sheathes with great care in constantly lengthening loops, the palpi affording much assistance in pushing the old skin downward. After the head and antennæ are entirely freed, the insect remains for a short time motionless, as-if to recover from its exertions. Very soon, however, it renews its efforts in a series of rapid jerks and contractions, by which the body is impelled forward while the outgrown skin is held firmly in place by the claws of the middle and posterior legs, which remain fixed in the wood. The most difficult part of the whole process seems to be the extrication of the front legs. This once accomplished, the

Katy-did has something to grasp with, and experiences no further trouble in withdrawing the body and the remaining legs from the old integument, often leaving the latter as an almost transparent shell in perfect shape upon the twig. It is not allowed to re main long, however, as an object of curiosity, for almost the first efforts of the transformed insect are directed to the task of eating up this, its out-grown and out-worn garment. When first out of its pupal covering, the wings of the mature insect hang down on each side as flexible and shapeless as strips of dampened lace; but, they soon begin to dry and harden, and are, by degrees, drawn up into place. The anterior pair, which were at first transparent, become gradually green and opaque, and display the characteristic leaf-like veinings; while the broad underwings, formed of transparent membrane intersected by an exquisite net-work of green veins, are folded fan-like beneath them, with only the tips for about a third of an inch visible, this portion being green and thickened like the wing-covers. The whole operation of moulting is performed within an hour."

ITS NATURAL ENEMIES.

Two birds aid materially in destroying the young Katy-dids. The Mocking Bird, (*Mimus polyglottus*,) and the Loggerhead Shrike or Butcher Bird, (*Collurio ludovicianus*.) Both build their nests in the orange tree or in live-oaks close by, and both have young in April. I have found Katy-dids impaled on orange thorns by the Shrike. I once saw a four-footed enemy, the Six-lined Lizard, (*Ameiva sex-lineata*,) with a full-grown one in its mouth; at first it had great difficulty in swallowing such a large morsel, but finally succeeded. The most important enemy, however, is a small Chalcid Fly, just bred from the eggs by Professor Riley, and also bred by myself. He calls it " The Black Rolling Wonder."

THE BLACK-ROLLING WONDER.

(*Antigaster mirabilis*, Walsh.)

ITS NATURAL HISTORY.

This curious little chalcid fly, averaging but from .09 to 14 of an inch in length, is the most inveterate foe of the Katy-did, consequently is of the greatest importance in diminishing its numbers. The female was first discovered by Walsh. See Vol. 2 of the "American Entomologist," pages 368–370, but the male and its life-history remained unknown until 1874, when Prof. C. V. Riley published a description in his "Sixth Annual Report on the Noxious, Beneficial and other Insects of the State of Missouri," page 162, from which I shall quote:

"The anomalous habit of this little fly consists in the habit possessed by the female of rolling up into a ball backwards, and in the very great dissimilarity of the male. Many other insects roll up downward, with a convex back, while some few, as the Rove-beetles, (*Staphylenidæ*,) curl up more or less backward when disturbed; but no other species is so curiously constructed for rolling backward into a perfect ball, unless it be some belonging to the very closely allied genus *Eupelmus*.

" These little parasites have always issued in the spring of the year, just about the time the young Katy-dids would have issued if they had not been molested; but as Mr. Walsh captured his specimens in August, the insect must either be double-brooded, or the female must survive during the summer months."

As I have bred it from sphinx eggs, the first is doubtless the correct hypothesis.

The larva of this little anomaly I have not yet met with, but the pupa is characteristically flattened and straightened to suit its narrow egg-abode. When mature, and not till the wings are expanded and all its parts are hardened, the fly gnaws its way out through an irregular, but usually round hole at the anterior or exposed end of the egg. The male, (Fig. 19,) as will be

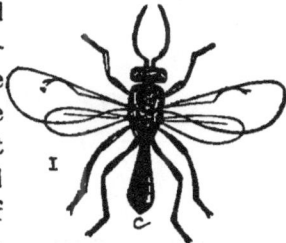

Fig. 19. After Riley.

seen by glancing at the figures, approaches much nearer the more common chalcididan form. He has clear wings, a narrow body, and is of a more brilliant metallic green color than the female, (Fig. 20,) from which, indeed, he differs so much that had I not bred both sexes from the same batch of eggs, I should scarcely believe them to be at all specifically connected.

In January, I put some of the eggs of the Katy-did into a tin-box, expecting to raise Katy-dids. I was much surprised,

Fig. 20.   After Riley.

therefore, when, the first week in April, on opening the box, I found a lot of these small flies. On examining my Katy-did eggs, a small round hole was plainly visible in the top of each, through which the parasite had come after its long winter feast. It had come forth to enjoy the sunshine, regale in the awakening beauties of spring, marry and perpetuate its kind.

RESUME.

In the fall, when the Katy-did lays her eggs, the female chalcid, ever on the watch, inserts her ovipositor into each one, at the same time depositing an egg. These, on hatching, begin to feed upon the albuminous substance contained therein, and like other ichneumon or chalcid flies, on arriving at maturity, change into pupæ and transform into flies, freeing themselves by eating round holes through their prison walls. I have raised these in the spring and fall, proving them to be double brooded.

REMEDY.

The best method of destroying the Katy-dids is to go over the trees in winter and spring, look for the eggs, and destroy them.

THE LUBBER GRASSHOPPER.

(*Rhomalea Microptera*, Burm.)

[Ord. ORTHOPTERA.   Fam., LOCUSTIDÆ.]

For some unaccountable reason in America the term "Grasshopper" has been given to an insect, which, in Europe, Asia and Africa is known as the Locust; while the latter term here is generally applied, or rather misapplied, to the different species of *Cicadas*. A mistake which, in America, is universal and will probably never be rectified.

Locusts, (grasshoppers) in Eastern and tropical countries have, from time immemorial, been used as an article of food. The wild Arabs and Bedouin tribes of the sandy deserts, the Hottentots and Kaffir tribes of Central Africa, the Indians of South America, and some few of the Indian tribes of North America, all enjoy the delectable dish of "fried grasshoppers." Even in the Bible, in the days of Christ, we read that John the Baptist lived on "Locust and wild honey." Centuries before, when Babylon and Nineveh were in their glory, we read that at feasts *Locusts* were considered a great delicacy.

Austin H. Layard, in his "Discoveries among the ruins of Nineveh and Babylon," published in 1853, gives the following interesting account of a sculptured slab, on which is represented servants carrying fruits, *Locusts*, pomegranates, &c., to a feast in the palace of Nebuchadnezzar:

"During my absence in the desert, the excavations at Konyunjik had been actually carried on under the superintendence of Toma Shishman. On my arrival he described many interesting discoveries, and I hastened to the ruins, crossing in a rude ferry-boat the river, now swollen by the spring rains to more than double its usual size.

"The earth had been completely removed from the sides of the gallery, on the walls of which had been portrayed the transport of the large stone and the winged bulls. An outlet was discovered near its western end, opening into a narrow, descending passage; an opening, it would appear, into the palace from the river side. Its length was ninety-six feet, its breadth not more than thirteen. The walls were paneled with sculptured slabs about six feet high. Those to the right, in descending, represent a procession of servants carrying fruit, flowers, game, and supplies for a banquet, preceded by mace bearers. The first servant following the guard bore an object which I should not hesitate to identify with the pine-apple, unless there were every reason to believe that the Assyrians were unacquainted with that fruit. The leaves sprouting from the top proved that it was not the cone of a pine tree or fir. After all, the sacred symbols held by the winged figures in the Assyrian sculptures may be the same fruit, and not, as I have conjectured, that of a coniferous tree. The attendants who followed carried clusters of ripe dates and flat baskets of osier-work, filled with pomegranates, apples and bunches of grapes. They raised in one hand small green boughs to drive away the flies; then came men bearing hares, partridges and *dried locusts fastened on rods.* The locust has ever been an article of food in the East, and is still sold in the markets of many towns in Arabia. Being introduced in this bas-relief amongst the choice delicacies of a banquet, it was *probably highly prized by the Assyrians.* (Page 289, N. Y. H. and Bros., 1853.

He also gives an account of the mode of preparing them, taken from "Burckhardt's Notes on the Bedouins, p. 369," and which I shall quote for the benefit of those who desire to try this ancient dish: "The Arabs, in preparing locusts as an article of food, threw them alive into boiling water, with which a good deal of salt has been mixed; after a few moments they are taken out and dried in the sun. The head, feet and wings are then torn off; the bodies are cleansed from the salt and perfectly dried, after which process, whole sacks are filled with them by the Bedouins. They are sometimes eaten boiled in butter, and they often constitute materials for a breakfast, when spread over unleavened bread mixed with butter "

It has been conjectured that the locust eaten by John the Baptist in the wilderness, was the fruit of a tree; but it is more probable that the prophet used a common article of food abounding even in the desert.

ITS NATURAL HISTORY.

As the name *microptera* indicates, it has very short wings, reaching only half way to the abdomen, and rendering but slight assistance in flight. It is very sluggish in its habits, and can easily be captured. "When full grown, it measures two and a half inches in length, and is of a yellowish color, barred and spotted with black. The wing covers are yellowish, shaded with rosy pink, and barred and spotted with black. The larvæ are shaped like the mature insects, but have not the rudiments of wings. They are of a black color, beautifully striped and banded with orange yellow. The pupæ have very small rudimentary wings, black, shaded and bordered on the thorax with yellow; the abdomen and hind thighs banded with same color. These insects destroy many garden vegetables and plants, and may be seen crawling sluggishly over the ground or upon shrubs, and are so nauseating that even the fowls rejected them for food. As they never fly, merely creeping or jumping heavily, they can easily be destroyed, by crushing with the foot, in every stage of their existence." (Glover.)

The eggs are laid in the ground during October and November, generally at the foot of an orange tree or close to a shrub, that, on hatching, the young may have abundant food. They hatch in the spring from the first to the middle of April, and may often be seen congregated around a small bush or young orange tree. On first hatching, they are dark brownish, nearly black, with a broad reddish or orange colored strip down the ridge of the back. Almost immediately after hatching, they climb to a green bush or up to the leaves of a young orange tree and begin to feed. They are vigorous and omnivorous feeders, being fond of nearly every green thing, particularly tender succulent leaves of young orange trees, which they frequently strip of all foliage.

REMEDY.

The best time for killing them is just after hatching, while they are clustered together and before they separate. At this time they are small, frail, soft creatures, and can be easily killed. No enemies to them have yet been discovered.

THE LARGE BLUISH-WHITE WEEVIL.

(*Pachnœus opalus*, Olivier.)

[Ord., COLEOPTERA. Fam., OTIORHYNCHIDÆ.]

BIBLIOGRAPHICAL.

Pachnœus opalus, Oliv. (Cucutis) Ent. v. 83, p. 339, pt. 24, fig. 345. Boh. Sch. Gen. Curc. vi 1, p. 425.

This is a large, oblong, oval weevil, of a bluish-white color, caused by minute powdery scales, averaging from .35 to .42 of an inch in length, having a moderately long snout with twelve longitudinal rows of slight punctures in the wing-covers. Abdomen bluish-white. The males are always the smallest.

ITS FOOD PLANTS.

This weevil was caught by me in great quantities in South Florida, on the Keys, feeding on the leaves of the Lime Tree, (*Citrus.*) I also found it eating the leaves of *Baccharis halim-*

ifolia and *Borrichia frutescens*, which I think are its natural food plants.

DESCRIPTIVE.

Form oblong, oval, densely covered with pale blue scales with a faint cupreous lustre. Body winged. Head sparsely punctured, densely scaly. Thorax broader at base than long, narrower in front, sides moderately arcuate, apex faintly lobed, base bisinuate, disc moderately convex, median line feebly impressed, surface densely scaly, median line and sides paler. Elytra densely scaly, with twelve rows of moderate punctures; the ninth somewhat confused, intervals indistinctly biserately punctulate. Body beneath densely scaly, scales larger and paler than above. Legs densely scaly, tibiæ, with short hairs on the inner side. Length .40 inch, 10 min.

Occurs in Florida and is not rare. The base of the elytra is not only bisinuate, but there is also a small dentiform prominence contiguous to the thoracic hind angles. Lacordaire mentions this character for two Cuban species, but not for our own. (LeConte and Horn the Rhynchophora, of America, north of Mexico, page 82).

———

THE SMALL BLUISH-WHITE WEEVIL.

(*Artipus floridanus*, Horn.)

[Ord., COLEOPTERA.   Fam., OTIORHYNCHIDÆ.]

This is another weevil, somewhat similar in form and color to the other, but much smaller, averaging but .24 to .26 of an inch in length. The thorax is unevenly punctured; instead of twelve longitudinal lines of punctures on the wing-covers, there are but ten, and these very unevenly deeper punctured, and punctures much larger than in the other species.

ITS FOOD PLANTS.

It was first sent to me by Mr. H. S. Williams, of Rockledge, Florida, who reported it eating the leaves of his orange trees. I also found it in my recent trip to the Florida Keys, feeding on the leaves of the Lime Tree, *Citrus Baccharis halimifolia* and *Borrichia frutescens*. Its habits and food are, therefore, similar to the other one. Unlike the other species, however, it was more abundant on *B. frutescens*.

REMEDY.

Both of the above insects may easily be removed by spreading a large sheet under the tree and gently shaking it; for, like all weevils, on disturbing, they loose their hold and immediately drop to the ground, seeking safety by hiding beneath rubbish under the tree.

DESCRIPTIVE.

Form oblong, surface densely clothed with white scales, varying to pale greenish blue, having cupreous lustre. Head and rostrum not as long as the thorax, sparsely punctured and densely scaly. Thorax as wide as long, cylindrical, slightly narrower in front, sides very slightly arcuate, apex and base truncate, disc moderately convex, median line moderately impressed, interrupted, surface densely scaly. Elytra nearly twice as wide as long, broadest behind the middle, sides feebly arcuate, base sub-truncate, disc moderately convex, feebly striate, striæ with moderate but very unequal punctures not very closely placed, intevals nearly flat, each with two rows of short scale-like hairs, surface densely scaly, the lower punctures surrounded by a darker area. Body beneath and legs densely scaly and sparsely hairy. Length 24 inches, 6 mm.

On examining the anterior tibiæ with rather high power, minute denticulations may be detected. This species resembles one from Cuba, (sent by Prof. Poey, without name,) which has the elytral intervals more convex, the punctures of the striæ larger, more regular and closer, and the thorax more densely punctured.

Several specimens from Key West. (LeConte and Horn, the Phynchopera, of America, north of Mexico.)

SUPPOSED ORANGE BORER.

(*Platypus compositus*, Say.)

[Ord., COLEOPTERA. Fam., PLATYPODIDÆ.]

BIBLIOGRAPHICAL.

Platypus compositus, Say, Journ. Acad. Nat. Sci., Philadelphia, vol. iii, 324. Do Am. Entom., Ed. by Leconte, vol. ii, 182. Er. Wiegm. Arch., 1836, vol. ii, 65. Chapuis Mon. Plat., 163, f. 75. *P. parallellus*, Chap. ibid, 164, f. 76. *Bostrictus parallellus*, Fab. Syst. El., vol. ii, 384. *P. tremiferus*, Chap. Mon. Plat., 174, f. 85. *P. perfossus*, Chap. ibid, 176, f. 86. *P. rugosus*, Chap. ibid, 176, f. 87. (Leconte, Horn, Rhynchophora, U. S.)

This is an elongated, cylindrical, reddish brown beetle, from .14 to .18 of an inch long, with the wing covers deeply striate, and ending in two sharp teeth at tip.

ITS DISTRIBUTION.

In "the Rynchophora of America," Dr. Leconte says it has been found from "Illinois to Texas, Louisiana, Florida, and South Carolina," so that it is pretty widely distributed.

The insect has always, I believe, confined its attacks to the Pine tree, into which it bores and lays its eggs, the larva boring into and feeding upon the wood. Specimens were brought me by Dr. O. J. Kenworthy from up the St. Johns River, said to have been found boring into orange trees, but I have never been able, authentically, to establish this fact. I have frequently heard of a *borer* in the orange tree, but have been unable to secure specimens in order to find out what it is. I am inclined to think this little beetle has been unjustly accused of doing the damage. A stray beetle might now and then, during its flight at night, settle upon an orange tree, and at daylight, naturally hide in any crevice.

THE ORANGE SPIDER.

(*Epeira*, Sp. ?)

In Moore's "Orange Culture," under the head of Injurious Insects, page 59, is the following account of this spider:

" One is a spider with a long slender body. When at rest, its fore-legs extend forward and the hind-legs backwards, and all parallel with the body, which clings closely to the branch or leaf on which the insect rests. In this position it would frequently be mistaken for a piece of moss or a rusty place on the bark It is so very timid that it at once attempts to conceal itself in this crouching position on the approach of any person. This position not only enables it often to elude observation, but generally to escape suspicion. I have watched it closely for two years, and was very slow to believe that so innocent looking a thing could have done the damage universally found in its presence. But I am fully satisfied that it is the cause of one of the forms of the diseases known as die-back.

" Early in the morning the insect is usually found on the tenderest shoots of the orange, and wherever found, the indications are the same. If the shoot is very young and tender, it begins at once to lose its freshness and ceases to grow; a little later, it assumes a rusty appearance and finally dies."

Now, I have given this spider special study. Mr. Emerson,

to whom I sent specimens, states that it is a species of *Epeira*, a *genus*, the species of which are difficult to determine, and its habits but little known.

Careful observations have failed to detect it damaging the orange tree; it is also contrary to the known habits of spiders to feed on vegetable juices. It is well known that die-back trees are constantly exuding sap from the tender, succulent shoots, which, in time, hardens. This sap attracts numerous small bugs and flies. It is therefore my opinion that these spiders collect in order to feed on these insects. I have caught eight or nine different species on orange trees. The female lays numerous pearly white eggs on the under part of an orange leaf, which she protects by weaving over them a fine silken web, straddling it in order to guard it from all enemies. She also shows solicitude and fondness for her eggs. These spiders are very timid, and on being disturbed, will try to escape, by dropping from the twig or leaf to which they are suspended by a silken thread.

THE ORANGE APHIS.

(*Siphonophora citrifolii*, N. Sp.)

[Ord., HEMIPTERA. Fam., APHIDIDÆ.]

Early in the spring and during the summer, until late in the fall, brownish or black plant-lice, as they are familiarly called, may be seen congregated in various stages of development in the tender shoots and branches of the oranges. These are the famed milch cows of the ants. If you watch them carefully, you will see that their beaks are inserted into the leaves or the tender succulent shoots, from which they extract the juices by a pumping motion of the body. The ants, clustering over them in crowds, carefully and gently touch the honey-tubes on the hinder part of the aphis with their antennæ. The aphis yield up their honey— a small drop oozing out at a time—which the ants eagerly lap up with their palpi. The matured aphids are black, with long seven-jointed antennæ and full, round, black abdomens. (Plate 1, Figs. 1 and 2.) The winged specimens are also black.

Winged individuals.—Head black, tubercles of antennæ stout and black, top of head reddish, rostrum long; thorax and abdo-

men black, shining; antennæ not quite reaching to end of abdomen, variable, marked somewhat like apterous females. Wings
hyaline, stigma rather broad, narrowing towards base and obliquely sharpening to point at outer edge towards apex, stigmal
vein curved sharply upwards, three oblique veins, the third
forked; hind wings with two oblique veins; *some specimens have
two in one wing and one in the other.*

ITS NATURAL ENEMIES.

The larvæ of two or three Syrphus and Tachina Flies, besides
the Blood-red Lady Bug, Twice-stabbed Lady Bug, the Orange
Chrysopa, Weeping Golden Eye, (*Chrysopa plorabunda*, and a
Freckled Chrysopa, (*Hemerobius*,) prey upon this species, destroying immense quantities. I have also bred two internal parasites,
a chalcid fly and an ichneumon fly, and from a larva I found
feeding upon them on the leaf, a curious fly. These I shall describe further on. This species is remarkable for its prolificness,
and besides laying eggs and producing alive, it is agamie, i. e.,
producing young without the intervention of the males. All
these causes help to swell their number and render them difficult
to exterminate.

THEIR DISTRIBUTION.

In all countries the Aphididæ are known in the vegetable garden, on shrubbery, fruit trees and other plants. Benj. D. Walsh,
in 1863, enumerated sixty-two different species; Riley & Monell,
in 1879, forty-nine species, west of the Mississippi; Prof. Cyrus
Thomas, in his report as entomologist of Illinois, enumerates one
hundred and sixty as occuring in the United States.

As I can find no description that will suit the species under
consideration, I·propose the name of *Citrifolii* for it.

DESCRIPTIVE.

SIPHONOPHORA CITRIFOLII, N. Sp. APTERUS FEMALES.—Length .05 to .06
of an inch, .03 to .04 in width. Either black and shining or dark brownish
black. Head between antennæ reddish. Antennæ, seven-jointed, pale yellowish, upper end of joints 3, 4, and 5, reddish brown, seventh joint thin, long and
setaceous. Legs pale yellowish, setaceous, latter two-thirds, of emoræ reddish
brown, end of tibiæ and claws reddish. Nectaries black, long and cylindrical.

REMEDY.

All tender shoots, badly affected should be removed and the
lice burnt or scalded. An application of powdered sulphur,

strong soap-suds, a wash made from tobacco leaves, or any strong lye wash may be used.

Professor Thomas says: "The most effective remedy, where it can be applied, is tobacco smoke or the fumes of burning tobacco, sulphur, &c. But to render these successful, the plants must be covered in some way, so as to confine the smoke or fumes and cause them to penetrate to all parts. A frame in the shape of a box or bell, covered with cheap cloth of any kind, will answer very well for the flower or vegetable garden, and might be used also for small bushes."

THE CHALCID FLY OF THE ORANGE APHID.

(*Stenomesius? aphidicola*, N. Sp.)

[Ord. HYMENOPTERA. Fam., CHALOIDIDÆ.]

This September I determined to find out what particular species of internal parasite it was that preyed upon our Orange Plant Lice, or Aphides. I put several of those that I perceived had been parasitized into separate boxes. About two weeks afterwards, on examining my boxes, I discovered several very minute black four-winged flies, (Fig. 21,) different from any I

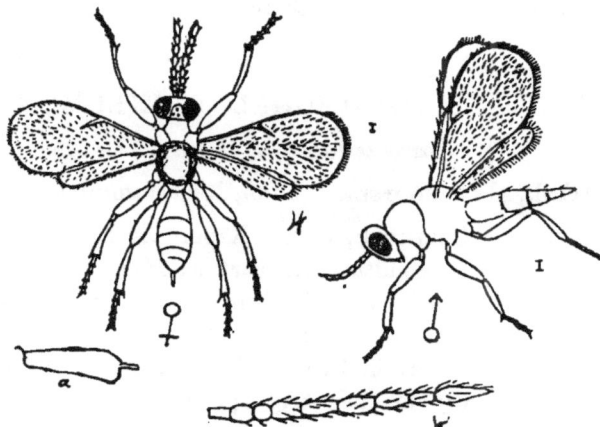

Fig. 21.

had ever seen or read of. On subjecting them to a microscopical examination, I found they were characterized as follows:

Deep black in color, finely punctured; females, nine-jointed antennæ, males 7 jointed, legs, excepting thighs, pale yellowish; thorax raised; abdomen of male long and slender; that of female broader and rounded; wings four, clear and veinless, fringed with short hairs, the fringing on hinder pair being slightly longer than on fore wings. This, then, is another little friend of the orange grower, with habits similar to those of the flies preying upon the scale insects. It differs, however, in this respect, the female, instead of laying a single egg, deposits two to three into the body of a single aphis. From a single aphis I raised three of these little flies.

As it is entirely different from any species known to me, I have provisionally placed it into Westwood's genus Stenomesius, until I can consult authorities. It is probably a new type of a genus.

DESCRIPTIVE.

STENOMESIUS? APHIDICOLA.—Female.—Length, .05 to .06 of an inch. Color, a deep black. Head wider than thorax,—three ocelli. Antennæ, nine-jointed, setaceous, inserted in front and close together. Scape short and narrow, second joint longer than third. Third, round, fourth, urn-shaped, joints 5, 6 and 7 subequal in length, 6th widest, 7th narrowest, 8th small, 9th twice as long as 8th. Abdomen as long but narrower than thorax, somewhat rounded, ending in ovipositor. Wings, hyaline, cell-less, shortly ciliated, ciliation longer on hinder than on fore wings. Legs, pale yellowish, femoræ and coxæ blackish or brownish black, tarsi five-jointed. Male.—Description the same as above, excepting slightly smaller, antennæ but seven-jointed, and not much longer than head is broad, while the abdomen is long, slender and pointed.

THE BLACK YELLOW-LEGGED ICHNEUMON FLY.

(*Trioxys testaceipes*, Cresson.)

[Ord., HYMENOPTERA. Fam., ICHNEUMONIDÆ.]

Prof. J. Henry Comstock, in his Report as Entomologist to the United States Agricultural Department for 1879, describes the discovery of another parasitic four-winged fly, belonging to the ichneumonidæ family. He says:

" The leaves of the terminal twigs of orange trees are frequently infested, especially in the early spring, by numerous dark-green plant-lice, which do considerable injury by checking the growth of the young shoots. At Rockledge, Fla., I found that these plant-lice were destroyed in great numbers by a small black ichneumon fly, a description of which has not heretofore been published. I bred the same species from plant-lice infesting the cotton plant, and from the common grain plant-louse, (*Aphis avenæ*.) The specimens were referred to Mr. E. T. Cresson, who prepared the following characterization of the species :"

Trioxys testaceipes, Cresson.—Female.—Piceous or shining black, smooth and polished, impunctured ; mandibles and palpi pale ; antennæ brownish-black, sometimes more or less pale beneath, thirteen-jointed, the joints faintly fluted or grooved, the last one longest and thickest ; wings hyaline, iridescent, stigma pale ; legs, including coxæ, yellowish, testaceous, the posterior pair generally more or less fuscous or blackish ; abdomen often brown or pale piceous with the first, and sometimes, part of the second segment, more or less testaceous. Length .07 of an inch. Hab.—Rocklege, Fla., Selma, Ala., and Pocomoke City, Md. Parasitic upon an aphid infesting twigs of orange, an aphid on the cotton-plant and *Aphis avenæ*, Fabr.

I have also raised a great many of this interesting species from parasitized aphis found on orange trees in my garden, and have had no difficulty in distinguishing it from Dr. Cresson's description. This not only proves it to be widely spread throughout Florida, but numerous, and therefore of the greatest importance in the destruction of the Aphididæ.

As Prof. Comstock and Dr. Cresson do not mention or describe the male, and as it differs somewhat from the female, I submit the following description:

DESCRIPTIVE.

MALE.—Length .06 of an inch. Antennæ, reddish-brown, twelve-jointed, last joint as long as ten and eleven combined, this joint being much longer in the male than it is in the female. Abdomen brownish, excepting first, second and third segments and the under surface, which are pale testaceous.

THE FOUR SPOTTED APHIS FLY.

(*Conops ? quadrimaculata*, N. Sp.

[Ord., DIPTERA. Fam., CONOPIDÆ.]

This curious looking fly was raised by me from a larva found feeding upon a cluster of Aphides on an orange leaf. The larva was of a flesh color, having a long neck and spiny-like tubercles along the back, with a reddish narrow band running down the centre. This changed into a puparium, (fig. 22,) of a dirty white, with brownish markings at the top, and very minute short hairy tubercles springing out from different segments. In about ten days, there issued from this a brownish wasp-like fly, (fig. 23,) with a bluish black head and short, three-jointed, club-shaped antennæ, apical joint furnished with a hairy spine; the wings clear, stigmal region slightly smoky; four yellowish white spots, two on second and two on third segments of abdomen. The eggs of this species are oblong and of a milky white color. The fly deposits two, three or more eggs in the midst of the plant lice; these hatch in from eight to ten days, and the larvæ begin to feed on the young aphis. They destroy countless numbers of Aphides during the season, frequently ridding the orange trees of these pests.

F. 22.

Fig. 23.

DESCRIPTIVE.

CONOPS? QUADRIMACULATA, N. SP.—Length .40 of an inch. Head much wider than thorax, bluish-black piceous, shining, three ocelli triangularly arranged and on a raised part of vortex. Antennæ clavate, short, three-jointed, brownish, darker at tip, apical joint longer than one and two and furnished with a single bristle, face, from antennæ down reddish piceous, with two oblong white downy bands, next the eye, extending from labrum and terminating just back of antennæ, also a very narrow line of white greyish hair back of eyes, extending from half the eyes downwards. Thorax longer than head is broad, black, finely punctured, covered with extremely short fine hair. Metathorax yellowish, brownish on top. Abdomen brown, wasp-like, clavate, more than twice the length of thorax and covered like thorax with hair. First joint long and narrow, joints two and three about equal with two oblique yellowish white spots extending from base, not joining on top, and running obliquely downwards not quite to half of the segment, fourth segment one-third the length of third, others very short, almost hidden in each other. Under surface much paler. Wings hyaline, slightly dusky, more perceptible about stigmal region. Legs reddish, anterior and middle pair darker at base of femora and tarsi, posterior pair, nearly the whole femora, excepting just before juncture with tibia dark brown, apical third of tibia and tarsi dark brownish.

THE APHIDIUS OF THE ORANGE APHIS.

(Aphidius? citraphis, N. Sp.)

[Ord., ICHNEUMONIDÆ. Fam., BRACONIDÆ.]

Another enemy of the orange aphis, which I raised this fall from specimens obtained from trees in my garden, may be called

"The Aphidius of the Orange Aphis." It proves to be a new species, and its description is as follows:

APHIDIUS CITRAPHIS, N. SP.—FEMALE.—Length .07 of an inch. Black, smooth, shining. Mandibles and palpi yellowish. Antennæ brownish, sixteen-jointed, first two joints equal, moniliform, closely joined, other joints longer, loosely joined and slightly decreasing in size, last joint being the smallest. Wings hyaline, veins and stigma yellowish, first cubital yellowish or partly so, varying. Legs honey yellow, hinder pair darker. Abdomen brownish, approaching black at posterior end. Under part of first and part of second segment, yellowish.

MALE.—Length .06 of an inch. Fifteen-jointed antennæ. Coxæ and femoræ of posterior legs blackish. First, second and hinder parts of third and fourth segments of abdomen, yellowish. In other respects it agrees with description of female.

THE COMMON PSOCUS.

(*Psocus venosus*, Burm.)

[Ord., NEUROPTERA. Fam., PSOCIDÆ.]

Psocus venosus, Burm., 11–778–10 Walk. Cat., 484–9. *Psocus magnius*, Walk. Cat., 484–10. *Psocus microphthalmus*, Ramb. Neur. 321–6. *Psocus aceris*, Fitch MSS., Collection of de Selys Longchamps. *Psocus gregarius*, Harris Cat. (Hagen Synopsis, 40–5.)

These interesting insects when full grown, average from .20 to .32 of an inch in length. They are of a dark brownish black, head brassy, with long thin antennæ covered with short fine hair; the fore wings are black, the three large principal veins yellowish, with a triangular yellowish white spot beyond the middle and near the margin towards the apex of anterior wings; legs pale yellow. Towards the winter months, these become quite numerous. It is both interesting and amusing to watch them congregated in flocks, from fifteen to forty, in all stages of development, with and without wings, crawling on fences, or up and down the trunks of trees, those with wings generally taking the lead.

Hagen, in his Synopsis of Neuroptera, states this species to be widely distributed, being found in New York, Washington, Ohio, Mexico, Cuba and Maryland.

ITS NATURAL HISTORY.

Like the Orange Psocus, it lays its eggs from eight to ten in number, under a web. I am unable to state how many one female deposits in a season, but judging from one which I killed and examined, I should say from seventy-five to one hundred. The eggs of this species are larger than the eggs of the other, and take about two weeks to hatch. Toward night, these little creatures all huddle together in a sheltered spot on the trunk of the tree, and here they remain until the sun comes out bright and warm the next day. I have often found them remaining together all through cold, damp, cloudy days. They may be found throughout the whole winter living in sheltered situations on trees, &c. The female hibernates during the winter, beginning to lay her eggs in March.

DESCRIPTIVE.

P. VENOSUS BURM.—FUSCOUS.—Head brassy. Antennæ blackish, fuscous, (in the male thicker, pilose,) the two basal articulations luteous. Thorax margined with yellow. The feet luteous, tarsi fuscous. Anterior wings fuscous, blackish fuscous, pterostigma triangular, yellowish, basil veins yellowish, apical ones fuscous. Posterior wings smoky, hyaline. Length to tip of wings 6-8 millimetres. Expanse of anterior wings 12-15 millimetres. (Hagen's Synopsis.)

THE ORANGE THRIPS.

(*Thrips*, Sp. ?)

(Ord., HEMIPTERA. Fam., THRIPIDÆ.)

In the flowers of the orange trees during February and March, are found numerous small insects which, under the microscope, present the following appearance:

Elongated and narrow, from .04 to .06 of an inch in length; beak, short and sharp; eyes large and prominent; antennæ long and covered with short hairs; the wings, four in number, (sometimes the hinder ones are aborted) are clear, narrow and bristly, placed wide apart, the front pair being much longer than hinder ones, and slightly widening at tip, which is rounded; the legs, six in number, are rather stout and short. This species is also very destructive to roses.

THE ANT-LION.

(*Myrmelion* Sp. ?)

[Ord., NEUROPTERA. Fam., HEMEROBIDÆ.]

In the soil under every orange tree, one often notices conical inverted shaped holes, made by what is popularly known here as the "doodle" or "ant-lion." These are caused by the larvæ of a large nerved winged insect, one of which lies concealed in the bottom of each hole. They have long curved jaws or mandibles, and the body is covered with long hair. Should an ant, bug, or other insect, fall into this pit, the strong curved jaws immediately close on it and the ant-lion then devours it at leisure. Sometimes a large species of ant falls into the trap; whenever this happens, a fierce battle ensues, the ant frequently coming out victorious.

This insect is beneficial in two ways—by destroying all injurious insects falling from the tree into the pit, and by burrowing in the soil around the roots, thus enabling the tree to gain air and moisture.

INDEX.